Radical Constructivism

Studies in Mathematics Education Series

Series Editor
Paul Ernest
School of Education
University of Exeter
Exeter

Studies in Mathematics Education Series: 6

Radical Constructivism:
A Way of Knowing and Learning

Ernst von Glasersfeld

UK   RoutledgeFalmer. 11 New Fetter Lane. London EC4P 4EE
USA   RoutledgeFalmer. 29 West 35th Street. New York NY 10001

First published in 1995
Reprinted 1996 and 1997

Reprinted 2002 by RoutledgeFalmer

Transferred to Digital Printing 2002

RoutledgeFalmer is an imprint of the Taylor & Francis Group

A catalogue record for this book is available from the British Library

Library of Congress Cataloging-in-Publication Data are available on request

ISBN 0 7507 0387 3 cased
ISBN 0 7507 0572 9 limp

Jacket design by Caroline Archer

Typeset in 10/12pt Bembo by
Graphicraft Typesetters Ltd, Hong Kong

Epigraph

The only given is
the way of taking.

Roland Barthes

Objectivity is the delusion that
observations could be made
without an observer.

Heinz von Foerster

Contents

Contents

Contents

Preface by Series Editor

Mathematics education is now established worldwide as a major area of study, with numerous dedicated journals and conferences serving national and international communities of scholars. Research in mathematics education is also becoming more theoretically orientated. Vigorous new perspectives are pervading it from disciplines and fields as diverse as psychology, philosophy, logic, sociology, anthropology, history, feminism, cognitive science, semiotics, hermeneutics, post-structuralism and post-modernism. The series *Studies in Mathematics Education* consists of research contributions to the field based on disciplined perspectives that link theory with practice. It is founded on the philosophy that theory is the practitioner's most powerful tool in understanding and changing practice. Whether the practice is mathematics teaching, teacher education, or educational research, the series will offer new perspectives to assist in clarifying and posing problems and to stimulate debate. The series *Studies in Mathematics Education* will encourage the development and dissemination of theoretical perspectives in mathematics education as well as their critical scrutiny. It aims to have a major impact on the development of mathematics education as a field of study into the twenty-first century.

In the past decade or two, the most important theoretical perspective to emerge in mathematics education has been that of constructivism. This burst onto the international scene at the exciting and controversial Eleventh International Conference on the Psychology of Mathematics Education in Montréal, in the Summer of 1983. No one who was there will forget Ernst von Glasersfeld's calm and authoritative plenary panel presentation on radical constructivism, and his replies to critics. That controversy confirmed his earlier observation that 'To introduce epistemological considerations into a discussion of education has always been dynamite' (Glasersfeld, 1983, p.41). Ironically, the attacks on radical constructivism at that conference, which were perhaps intended to fatally expose its weaknesses, served as a platform from which it was launched to widespread international acceptance and approbation.

In this volume Ernst von Glasersfeld offers what I believe to be the definitive theoretical account of radical constructivism. It is an elegantly written and thoroughly argued account of this epistemological position, providing a profound analysis of its central concepts. Although he indicates his debt to Jean Piaget (and indeed to collaborators such as Leslie P. Steffe), Glasersfeld shows that the roots of radical constructivism are much older. A great strength of the book consists in the two genealogies of knowledge which are offered as an orientating basis. These veritable genetic epistemologies trace the development of the central ideas of radical constructivism along two tracks. The first is the history of philosophy from the pre-Socratic masters of Ancient Greece to the present. The second is his own intellectual biography. In it Glasersfeld illustrates how a number of lines of thought from cybernetics, linguistics, developmental psychology, cognitive science and philosophy became synthesized into radical constructivism. Given these diverse roots, I expect this first full articulation of the theory to have an influence that extends beyond mathematics education.

Radical constructivism is a progressive research programme with many strengths. To mention but two, it is first of all a sceptical position in epistemology, which incorporates a fallibilist view of mathematics. This is consistent with much of the latest work in the philosophy of mathematics, as earlier volumes in the series show. Secondly, radical constructivism continues to grow and develop. The definitive account that this book provides will in no way inhibit its continued growth and development, and its practical applications. I predict that the book will be very influential in both grounding and stimulating further work in this orientation, and will be much cited by researchers investigating understanding and the learning of mathematics. I can think of no better volume to exemplify the series philosophy than the present one.

Paul Ernest
University of Exeter
July 1994

Reference

Glasersfeld, E. von (1983) 'Learning as a Constructive Activity', *Proceedings of Fifth Annual Meeting of the North American Chapter of the International Group for the Psychology of Mathematics Education*, Vol. 1, Montréal, PME-NA.

Preface

Twenty years ago, when Charles Smock and I put together our research report on epistemology and education, we chose as subtitle: 'The implications of radical constructivism for knowledge acquisition' (1974). It was the first time that the word 'radical' was associated with Piaget's genetic epistemology. Charles, who had worked with Piaget at Geneva, sent a copy of the report to the master, who had introduced the constructivist approach to cognition in the 1930s. A few weeks later, Charles received a most encouraging acknowledgment: 'I always appreciate what you write,' Piaget said, 'you are one of the few Americans who have understood me' (April, 1975).

Since then I have learned that Piaget was a most reluctant reader of other people's writings. In the case of our report, I obviously prefer to think that he looked at it.

About that time I began to work with Les Steffe on the constructivist approach to the learning and teaching of arithmetic. Without him, radical constructivism would have remained a private enterprise. His flair for producing plausible operational analyses of what elementary-school children seem to be doing when they try to handle numbers, led to practical applications in school rooms. Wherever the experiment was continued for at least two years, the results far exceeded our expectations. More than anything else, this encouraged me to continue with the elaboration of the constructivist theory of knowing.

What we did not expect at all, was that 'radical constructivism' would become a catch word — with all the advantages and disadvantages popularization brings. Reactions have varied a great deal, and on both the positive and the negative side, they have at times been somewhat passionate. The purpose of this book, therefore, is to lay out the main constructivist ideas as I see them.

It will surprise some readers that I occasionally pit my ideas against behaviourism. They may feel that I am flogging a dead horse. I would agree that behaviourism is *passé* as a movement, but some of its central notions are still very much alive, both in psychology and education. Those who cling to them are likely to get a distorted view of constructivism.

Most of this text is new, but the ideas it expounds have been central to my work for many years. Some are expanded here, others compressed. Wherever there are actual overlaps with earlier papers, I have indicated them.

In the first chapter I recount how I personally came to adopt a constructivist way of thinking. The second is a biased run through the history of western philosophy. It is intended to show that there is nothing very new about the ideas that form the core of my orientation. I limited the survey to philosophers and did not mention writers, such as Pirandello, Musil, and Fowles, who developed their own constructivism. Even so, I am aware of holes, and have no doubt that there are omissions of which I am not aware.

I suspect it happens to everyone who is trying to formulate the results of lengthy reflections. You sit with pen in hand, or at your keyboard, and look at the sentence you have just written. There seems to be nothing you want to change. But suddenly you feel uncertain: where did this come from? Have I read it somewhere? You search your mind and ask yourself, who might have said such a thing?

I have a lot of books, and on some occasions I spend half a day or more looking in likely places, usually without success. It is not easy to spot a sentence in hundreds of pages, especially when the search must be in different languages. And if one has done some reading in libraries that are now far away, it is impossible to retrieve forgotten sources. By now, I have given up. I realized that whatever I write will have been said by someone somewhere. Ideas, I tell myself, should never be personal property. What matters is how one uses them. Thus I have tried to give credit wherever I can, and hope not to have made improper use of what I may have unknowingly appropriated.

Chapters 3, 4, and 5, explain my reading of Piaget and how some of the ideas I had used in my earlier work on conceptual analysis could be merged with his. Chapters 6, 7, and 8, present models that to me seem crucial elements in the constructivist edifice: the concept of self and others, the function of language, and the view of information and control that was developed in the field of cybernetics.

Chapter 9 describes a hypothetical model for the generation of units, pluralities, and numbers. It is the model that Steffe and I have been using in our work on the development of arithmetical concepts in children. Although these concepts are very specific, I feel that the method of analysis could be useful to teachers, educators, and, more generally, anyone who is interested in developing or modifying the conceptual structures that others seem to have in their heads.

Chapter 10 was the most difficult to write, because in the field of education I feel more of a foreigner than in the other disciplines I have invaded. It is, again, in the personal genre of Chapter 1, and reflects my view of education. Most of the suggestions I make, are not my invention but stem from the ongoing work of others. I am painfully aware of having left out references to many who have successfully applied constructivist ideas in their research and teaching. I can only hope that the tentative survey I present will enable readers to recognize what I would call genuine constructivism when they find it in the contemporary literature.

Acknowledgments

All I have done and written was driven by the wish to acquire an attitude my parents had made me see: to follow clear thinking rather than dogmas, and to be loyal to people rather than nations. Their example was a wonderful gift.

As to the present, I want to acknowledge the loving support and encouragement I got from Charlotte, my wife. Her help should be appreciated by the readers of this book, because it was she who frequently reminded me of Wittgenstein's famous precept 'Everything that can be said, can be said clearly'; and it was she who saw to it that my sentences became shorter.

I am deeply indebted to the University of Georgia for having offered me, twenty-five years ago, an academic position though I did not have the usual qualification of a Ph.D. Working with colleagues in the Department of Psychology widened my horizon, and the intensive interaction with graduate students was invaluable in refining my ideas.

I thank Jack Lochhead for having offered me a place in his Institute after I retired from Georgia, for his unwavering friendship and support, and for the many times he pointed out conceptual jumps in my writings.

I also want to thank all those who took my work seriously and wrote about it with understanding — especially Siegfried Schmidt in the German-speaking world, Felice Accame in Italy, Jean-Louis Le Moigne in France, and Jacques Désautels and Marie Larochelle in Canada. That I have benefited from stimulation, help, and support from many others should become clear in Chapter 1.

With regard to this book, I am grateful to Paul Ernest and Falmer Press for having suggested that I write it. I have tried to live up to their expectations. If I were a little younger, I would take another year to iron out the creases. As it is, I hope it will stimulate readers to think along the lines suggested and to plug the gaps I am sure they will find.

Scientific Reasoning Research Institute,
Amherst, June 1994
E.v.G.

List of Figures

Chapter 1

Growing up Constructivist:
Languages and Thoughtful People

What is radical constructivism? It is an unconventional approach to the problems of knowledge and knowing. It starts from the assumption that knowledge, no matter how it be defined, is in the heads of persons, and that the thinking subject has no alternative but to construct what he or she knows on the basis of his or her own experience. What we make of experience constitutes the only world we consciously live in. It can be sorted into many kinds, such as things, self, others, and so on. But all kinds of experience are essentially subjective, and though I may find reasons to believe that my experience may not be unlike yours, I have no way of knowing that it is the same. The experience and interpretation of language are no exception.

Taken seriously, this is a profoundly shocking view. Some critics say that the emphasis on subjectivity is tantamount to solipsism (the view that nothing exists outside peoples' heads), because, they seem to think, it implies that individuals are free to construct whatever realities they like; others claim that the constructivist approach is absurd, because it disregards the role of society and social interaction in the development of an individual's knowledge. Both objections are unwarranted, and the later sections of this book will present formal arguments to demonstrate it.

I have mentioned the feature of subjectivity here at the outset, because I believe that the best way of providing an introduction to radical constructivism is to tell how I, as an individual subject, came to embrace it as a general orientation.

The beginning of this story, inevitably, has to do with my life and the roots of my dissatisfaction with traditional theories of knowledge. It will be a chronicle of gathering ideas from people I met and authors I read, none of whom, I suspect, would wholly agree with how I interpreted them and built up my model. Hence I want to preface my account with two explicit warnings.

The first is that everything expressed in this book is simply this author's view. It is an attempt to explain a way of thinking and makes no claim to describe an independent reality. That is why I prefer to call it an approach to or a theory of *knowing*. Though I have used them in the past, I now try to avoid the terms 'epistemology' or 'theory of knowledge' for constructivism, because they tend to imply the traditional scenario according to which novice subjects are born into a ready-made world, which they must try to discover and 'represent' to themselves. From the constructivist point of view, the subject cannot transcend the limits of individual experience. This condition, however, by no means eliminates the influence and the shaping effects of social interaction.

The second warning concerns my memories and the act of remembering generally. As the Italian philosopher Giambattista Vico (1744–1961) remarked, we cannot reconstruct the past exactly as it was, because we cannot avoid framing and understanding our recollections in terms of the concepts we have at present. Independently, two centuries later, Jean Piaget came to the same conclusion (1968). The story I am going to tell of my journey to constructivism, therefore, is the story as I see it now.

Which Language Tells It 'as It Is'?

Problems with the notion of reality cropped up early in my life because I grew up with more than one language. My parents were Austrians, and at home they normally spoke German. But up to the end of the first World War my father had been in the diplomatic service and he and my mother got very used to speaking English. When I was little, they would switch to English whenever they wanted to speak of things they thought unsuitable for a child — and there were more things of that kind then than now. For me, of course, this was a powerful incentive to get into that secret language as quickly as possible, and when I started repeating words and phrases I picked up from them, they could not resist correcting my imitations and helping me to learn English. As a result, I felt pretty much at home in both languages by the time I was about six years old.

When Czechoslovakia was created as an independent state after the first World War, my father, whose original home and property were in Prague, automatically became a Czech citizen (and so did my mother and I). He no longer could nor wanted to be an Austrian diplomat. He devoted himself to photography and settled in the South Tyrol, the part of Austria that had become Italy after 1918. There I occasionally played with Italian children and the elementary school I was eventually sent to, was half German, half Italian. Without trouble and almost without noticing it, I learned a third language. Because my mother was a great skier and mostly took me along into the mountains and when she competed in races, I effortlessly learned to ski and it became very much part of my early life.

At the age of ten, I was sent to Zuoz, a very international boarding school in Switzerland, where I got daily practice in all three languages. For the next eight years I was also taught French. Slowly I began to realize that learning a language in a school room was a different thing from learning a language in your every-day environment. The teacher explained what the

French words meant in the language we all knew (which happened to be German). He showed us how to pronounce them and explained the rules of grammar we needed to make sentences. Then we read French texts and learned to translate them. How we understood the words and the texts, therefore, was largely in terms of experiences we had had elsewhere. Thus the French we learned was grafted on the language we had grown up with. This was quite different from growing into a language by interacting with people who live in it day in and day out.

When we started to read Balzac and Maupassant and Anatole France, it dawned on us — perhaps because our teacher was a master at circumscribing what could not be translated — that to get into another language required something beyond merely learning a different vocabulary and a different grammar. It required another way of seeing, feeling, and ultimately another way of conceptualizing experience.

This was no more than a dim notion then, but it persisted because, after I had graduated from high school, it helped to make the multilingual world in which I lived a good deal more intelligible and congenial. Ingenuously and certainly without formulating it, I had stumbled on a way of thinking which, as I discovered some twenty years later, was the core of the well-known Sapir-Whorf hypothesis. Put in the simplest way, this hypothesis states that how people see and speak of their world is to a large extent determined by their mother tongue (Whorf, 1956). In retrospect, I think, it was my first-hand experience of this phenomenon that prompted my interest in epistemology. If language had something to do with the structure of my experience and therefore to some extent with the world that I considered to be real, I could not for long avoid asking the question, what the *real* reality behind my languages might be like and how one could know and describe it.

The Wrong Time in Vienna

From high school I went on to study mathematics, which I had always liked — perhaps because it seemed the only subject that did not depend on a natural language. I entered Zürich University, but after one semester my father told me that Swiss Francs were no longer available and if I wanted to continue my studies it would have to be in Vienna. I was not enthusiastic about this move, but I went there in the autumn of 1936. The Austrian Nazi movement, although officially forbidden, was making itself felt everywhere, including the corridors and lecture halls of the universities. It was a depressing atmosphere, and when, before the end of the second semester, I was offered the opportunity of a winter in Australia as ski instructor, I jumped at it.

As it turned out, this was to be the end of my academic education. But Vienna had introduced me to two authors that influenced me profoundly: Freud and Wittgenstein. Freud's work (especially his *Interpretation of Dreams*, 8th edition, 1930) suggested that one could try to devise a rational model of the workings of individual minds; and his method required that the analysis of what individuals had unconsciously implemented in their own minds always had to be brought forth by the individuals themselves. (Freud himself seems to have forgotten this principle in some later writings and many professional psychoanalysts disregarded it altogether.)

Wittgenstein's *Tractatus* (2nd printing, 1933) captivated me above all because of the elegant neatness of his exposition. It seemed convincing, even if I did not altogether understand it. During the years that followed I reread the book several times, and one day woke from the spell, when I came to proposition 2.223:

> In order to discover whether the picture is true or false we must compare it with reality. (Wittgenstein, 1933)

It suddenly struck me that this comparison was not possible. In order to make it, one needed to have direct access to a reality that lay beyond one's experience and remained untouched by one's 'pictures' and their linguistic formulations. I felt that there were things one could say and believe to be true in one language, and yet one could not translate them into another. There seemed to be no way of showing their truth outside the context of experience in the particular language. I put away the *Tractatus*, and began to look elsewhere. (Many years later, when I read it once more, I realized that there was much in it that I had not understood.)

Growing Roots in Dublin

A few months before Hitler started the war, Isabel, my wife, and I were in Dublin. She was British, and when my Czech passport expired, I was able to obtain a 'stateless' alien's permit. This allowed us to stay. It did not make me eligible for a regular job, but I could do freelance work or work on a farm. With a friend I had made in Dublin we invested what money we had in a small farm in County Wicklow. He had been trained as a farmer and I was fit enough to do field work. Walking behind horses with a plough or a harrow was wonderful work for me: most of the day it leaves you free to think.

After Wittgenstein, I had read Joad's *Guide to Philosophy* (1936) and some Bertrand Russell, and there was lots to think about. And then I had the good fortune to make friends with Gordon Glenavy and Ned Sheehy, two amateur philosophers who had studied Berkeley and interpreted him in a way that made very good sense to me.

Berkeley's famous dictum *esse est percipi* has usually been taken as an ontological statement, i.e., a statement about the nature of reality. According to this view, he was saying that being perceived generates the existence of things. If this had been his intention, the many quips made by his critics would be quite justified and one could conclude that it was indeed absurd to hold that a tree in the forest falls and makes a noise only if someone sees and hears it fall. But there are reasons to believe that he did not intend to say this. First of all, he wrote the Latin phrase at the very beginning (paragraph 3 of 156) of a treatise to which he gave the title: *Of the Principles of Human Knowledge* (1710). He was not a sloppy author who picked his titles thoughtlessly. If he chose this one, he intended to write about human knowledge, not about ontology. Second, he explicitly laid out what he meant by the Latin slogan:

> The table I write on I say exists, that is, I see and feel it; and if I were out of my study I should say it existed — meaning thereby that if I was in my study I might perceive it, or that some other spirit actually does perceive it. (Berkeley, 1710)

And Berkeley adds a general explanation:

> There was an odour, that is, it was smelt; there was a sound, that is, it was heard; a colour or figure, and it was perceived by sight or touch. This is all that I can understand by these and the like expressions. For as to what is said of the absolute existence of unthinking things without any relation to their being perceived, that seems perfectly unintelligible. (ibid., Part I, par.3)

He is, in fact, defining the way he, Berkeley, wants to use the words *esse* (to be), 'to exist', and 'existence', when he is concerned with human knowledge. He also asserts that, for him, the term 'existence' has no intelligible meaning beyond the domain of experience.

His ontology is a different matter. He was a believing Christian (so much so that he became a bishop) and he therefore based his ontology on revelation, not on rational knowledge. To make it jibe with his theory of knowing, he added a mystical detail: because God perceives all things all the time, their permanence is assured. But this permanence belongs to the domain of metaphysics, not to the study of rational human knowledge. (There is a great deal more I gathered from Berkeley throughout the years and his name will crop up in later sections of this book.)

In 1939 *Finnegans Wake* was published and, although Joyce had lived in self-imposed exile for about two decades, the event was celebrated like no other in Dublin's intellectual circles. An informal group was formed of people who knew other languages, to try and unravel some of the countless puns that make up Joyce's extraordinary text. The group lasted through two meetings during which we covered the first three pages, but then our enthusiasm as well as the supply of Irish Whiskey dried up. In the opening lines of *Finnegans Wake*, however, there is the first of many oblique references to Vico, a name I had never heard before. I was told that Giambattista Vico was an eighteenth-century philosopher in Naples. If Joyce had chosen him as one of the threads in his work, I thought, he must be worth reading. I discovered that the Public Library in Dublin had an early Italian edition of Vico's *Principi di scienza nuova* (1744), and the next few times I could steal away from the farm, I went there to read it.

Vico's notions that we can rationally know only what we ourselves have made, and that the knowledge of poets and myth-makers is of a different kind, fitted well between some of the disconnected ideas in my head. Only very much later did I come to read Vico's treatise on epistemology (1710), which, as far as I know, is a first explicit formulation of constructivism (see Chapter 2).

Interdisciplinary Education

Shortly after the war ended, my farming friend fell in love with a British visitor and wanted to follow her to England. Isabel and I had loved Dublin and the life on the land, but the desire to return to a drier, sunnier climate had grown as we developed the first symptoms of rheumatism. We sold the farm, and after I had acquired Irish citizenship, I once again had a valid passport and was free to travel. We managed to start up my old car that had been mouldering in a shed for six years, packed our books and our two-year old daughter, and left for Paris, Switzerland, and eventually Merano, my former home in Northern Italy. Though we had planned to return to Ireland, we postponed it when, by a fortunate accident, I met Silvio Ceccato. This meeting, more than any other event, determined the future course of my life.

Ceccato had studied music, composed an opera that had been performed, and become intensely interested in the literature on aesthetics. He found no illuminating answers and went on to spend some twenty years reading philosophy. When we met, he had concluded that there was something basically wrong with the traditional approach and that a different way could be found. He had a fairly clear idea what this way would be, and he assembled a group of interested friends to work it out. He called the group 'The Italian Operationist School' and they were then in the process of developing a new theory of semantics. The group comprised a logician, a linguist, a psychologist, a physicist, an engineer, and one of the first computer buffs in Italy.[1] None of them was fluent in anything but Italian, and when Ceccato discovered that I spoke four languages and had congenial interests, he asked me to join the group. Since he had explained that they were trying to reduce all linguistic meaning, not to other words, but to 'mental operations', I was immediately hooked.

The idea to define concepts in terms of operations stemmed from Percy Bridgman, the American Nobel-prize physicist, who had developed it in the context of analysing key-concepts in Einstein's theory of relativity (Bridgman, 1927). Unfortunately, Bridgman's 'Operationalism' was appropriated by the behaviourist movement in psychology and criticized by philosophers on the basis of excerpts that focused on the physical operations of measurement. What Bridgman had said about the mental construction of concepts was generally disregarded. For me, the thesis that words stand for concepts and that definitions should specify the operations one has to carry out to build up these concepts, fitted nicely with Vico's principle of the construction of knowledge.

During the following years, my apprenticeship in Ceccato's group, which met informally two or three times a year for a few days of intensive discussion, taught me to question all conventional ideas and the tacit assumptions in the traditional theories of knowledge. In 1949 Ceccato founded *Methodos*, an international journal on language analysis and logic, and I was asked to translate Italian and German contributions into English.[2] The pay was miserable, but it was a unique opportunity and I was able to earn most of my living as a journalist.

When Ceccato gave me his article for the first issue of *Methodos* to translate into English, I had no idea how difficult this would be. He had written a parody that presented the history of epistemology as a game, not unlike poker, in which the great philosophers of the western world were the players. The goal was to establish a fundamental value, but it was forbidden to agree on it at the outset. Each player, therefore, had to introduce his own choice surreptitiously and, if he was skilful, make it seem necessary and self-evident at the end (Ceccato, 1949).

Today, I cannot read my translation without embarrassment: there were allusions I did not understand and much of the irony passed me by. In time, however, the continuous contact with the journal widened my philosophical horizon and translating was the best possible training in the use of words.

A Close Look at Meanings

In 1955 Colin Cherry invited Ceccato to the third London Symposium on Information Theory and encouraged him to apply the results of his operational analyses to machine translation, a field of research that had recently sprung up. At the time the United States military commands were said to employ a large army of translators to keep up with scientific, technical and other information that was published in Russia, and they hoped machines would help to cut the waiting time.

By then Ceccato was lecturer in philosophy at the University of Milan. It was said that he received the appointment because some of his publications provided arguments against communism. However, when someone realized that the same arguments could be used against any dogma, including that of the Church, students were no longer advised to go to his lectures. The lectureship, however, enabled Ceccato to create the first Centre for Cybernetics within the framework of the University, and a proposal for research on machine translation could be submitted to the US Air Force (Ceccato, 1960). The proposal was accepted and, for the first time, Ceccato was able to pay people to work with him. In 1959 I became a full-time research assistant at the centre.

7

My first major task was to provide an analysis of the concepts that English and other languages, including Russian, express by means of prepositions (I had two native speakers of Russian to work with). It was a bottomless subject and it occupied me long after the Milan project had come to an end (Glasersfeld, 1965). To begin to see the complexity of the problem, one need only ask, for instance, what conceptual relation, say, the word 'by' indicates in the following expressions:

> He opened the box *by* brute force;
> She spent an hour *by* the river;
> This time we came *by* the fields;
> I tried to read her letter *by* moonlight;
> Have this ready *by* Friday!
> My doctor swears *by* vitamin C.

And there are other relations that had to be distinguished in a detailed analysis, because in each of the languages that concerned us one needed a variety of expressions to translate the English 'by'. Since this is a question of conceptual relations, it demonstrates that different languages determine different conceptualizations.

Working in this area (in which there are countless examples of conceptual discrepancies between nouns, verbs, and adjectives that are given as equivalent in bilingual dictionaries) confirmed my deep feeling that each language entails a conceptually different world. Translation, in the sense of rendering in the second language the identical conceptual structure that was expressed in the first, was impossible, and our conceptual analyses demonstrated why.

Of course, there is a great deal of practical overlap because the differences are often very subtle and seem irrelevant in ordinary experiential situations. What we call communication works well enough whether an English girl says 'I like that boy', or an Italian '*questo ragazzo mi piace*' — it does not seem to matter that the one expression assigns the active role to the girl, the other to the boy. But it does show that the speakers' worlds are conceptualized differently.

The American Connection

After Ceccato's project had come to an end, another US Air Force research office became interested in the type of conceptual analysis we had been doing and decided to finance a more modest effort which I directed at the Milan Institute for Documentation. They desperately needed translators and shared the hope that computers would soon be able to help with that job. Our project monitor was Rowena Swanson, and it was she who first brought me into contact with Warren McCulloch, Heinz von Foerster, and Gordon Pask, the leaders in the new field of cybernetics. Though Dr Swanson herself was not a scientist by training, she had the most remarkable understanding of the process of scientific research and the value of interdisciplinary connections. Her office became something of a clearing house for novel ideas and her policy of bringing together the people whose work she sponsored provided invaluable stimulation to everyone concerned.

In the following two years Piero Pisani, Jehane Barton, and I worked out a novel approach to the analysis of the meaning of sentences by computer (Glasersfeld and Barton Burns, 1962). Because the large machines in those days were few and rarely accessible, we represented our system on some 120 square feet of plywood on the wall of our office and simulated the computer's basic operations (reading, comparing, and writing symbols) by moving an army of drawing pins by hand. It was an incredibly slow procedure but had the advantage of making immediately visible any inconsistencies or bugs in the programme we were designing.

Then, in 1965, the Washington office told us that there would be no more money for our kind of research in Italy, but they would continue to finance us if we came to the United States. It was a difficult choice. None of my colleagues had been to America, nor had they ever considered leaving Europe. But we did want to go on with the project, and in the end we decided to make the jump. We arrived in Athens, Georgia, towards the end of 1966.

In the spring of 1969, Isabel, with whom I had shared life for more than three decades, died of an embolism, suddenly, without the slightest warning. It took me several days to grasp that the world I had taken for granted was gone. I began to realize that the notion that we construct our reality is not just an intellectual gambit. The reality we had built and sustained together was falling apart without her. The only thing that kept me going during the year that followed was work. And then Charlotte came along, agreed to marry me, and bit by bit we set out to build a new world.

And then there was another unexpected break with the past. At the end of 1969, Mr Nixon wiped out the Air Force Office that had been sponsoring us as well as some twenty other research projects in computer science and communication. It was then that, like the Good Fairy, the University of Georgia, where we had a contract with the Computer Centre, stepped in and adopted all of us. Brian Dutton, who had replaced Jehane Barton and whose Ph.D was actually in medieval Spanish poetry, slipped into the Romance Languages Department; Piero Pisani was snapped up by the Computer Center; and I was taken in by the Department of Psychology. So began a never contemplated life in academia.

Introduction to Psychology

Two members of the Department, who differed from their largely behaviourist colleagues, had some sympathy for my ideas about language and most generously helped me by letting me sit in on their courses. One was Bob Pollack, who just then (1969) had edited a book of Alfred Binet's research papers that showed the French author as a psychologist of far greater depth than the 'Binet Scale of Intelligence' might lead one to believe; the other was Charles Smock, a developmentalist who had been trained for some time in Geneva.

To keep up with Pollack's course, I had to read a lot on perception, because this was his specialty. As I had never thought about the mechanics of seeing, I learned a lot about the models current in psychology. On the one hand, I found them fascinating because of the ingenious experiments that provided the data with which the models could be 'confirmed'. On the other, I was amazed at the general lack of epistemological considerations. What the eye sees — light, colour, and shape — was usually taken for granted as a physical given, and the research focused on the sensory mechanisms that could convey a presumed reality to the brain. No one seemed to doubt the assumption that Wittgenstein had expressed so succinctly in his proposition 2.223 (see above). The aim of the experiments was always to discover how the eye manages to see what is there, as though to perceive were simply to *receive* something that exists ready-made. The naive metaphor of the photographic camera seemed to dominate the field, in spite of the fact that the scene in front of a camera, as well as the picture that comes out of it, are obviously a product of the very perceptual processes they were studying.

I did, however, come across one spectacular exception: the perceptual oddity experts occasionally referred to as the 'Cocktail Party Effect'. This is a phenomenon we all are familiar with, without having studied psychology. It can happen anywhere. You have been buttonholed by someone who is telling a boring story. Suddenly you become aware of a much more interesting conversation that is going on behind you. You don't want to offend the bore, so you follow what he says, but just enough to be able to make an encouraging noise whenever he pauses to catch his breath. The main part of your attention is on what is being said behind you. This means that you are able to switch your attention at will to different points in your auditory field. It is not a question of one stimulus being more 'salient' than another, because the speech of your boring companion is louder and clearer than the dim conversation of the people you don't see. It is obviously a question of your subjective interest.

I was fascinated by this and discovered that famous experimental psychologists, such as Donald Hebb, Karl Lashley, Wolfgang Köhler, and the Russians Zinchenko and Vergiles, had independently noticed and experimentally demonstrated the same phenomenon in the visual field. It struck me as a truly revolutionary fact, yet none of the psychology textbooks that I came to see during the following years mentioned it. I shall return to the mobility of attention later (see Chapter 9), but here is how Köhler (1951) described one of his results:

When two objects are given simultaneously in different places while the eyes do not move, we can compare these objects, and say whether they have the same shape. (Köhler, 1951, p.96)

In other words, we can see objects that are in different parts of our visual field and, *without moving our eyes*, compare one with the other. Our attention obviously has the power to move within the visual field just as it can move among and select from speech sounds that arrive at the same time. Indeed, in the visual field we do this quite often, for instance when our eyes are fixed on the computer screen and at the same time we notice that someone we know has walked past the window beside our desk.

For me the realization of this capability was an enormous encouragement to pursue the search for the active element in the perceiver and, ultimately, the builder of knowledge.

Collaboration with a Chimpanzee

That the University of Georgia adopted me as a psychologist was due to the fact that my interest and work in computational linguistics happened to fit into an empty slot in that department. Once more a lucky coincidence with this move led me into a venture that was as fascinating as it was unexpected. Ray Carpenter, one of the leading primatologists in the United States had joint appointments at the University of Georgia and the Yerkes Primate Research Center in Atlanta. Just at that time, the first reports on the chimpanzee Washoe, who was learning sign language, had been published by the Gardners at the University of Nevada. The discussion about whether or not a chimpanzee could acquire a language became heated and spread far beyond the specialized journals. The Yerkes Institute wanted to join the fray and planned to set up experiments that could provide more rigorous tests than the subjective evaluation of fleeting exchanges of hand signs. Ray Carpenter promoted the idea of creating a communication system consisting of keyboards and a computer that could record all interactions. When he heard of our work in computerized language analysis, he asked me whether I would be interested in designing the language and computer components of the system planned at the Yerkes Center. I talked it over with my computer colleague Piero Pisani, and we decided to go ahead.

I designed the 'Yerkish' language, using geometric designs as symbols for words (concepts) and a simplified but very strict grammar to govern their formation into sentences. By sequentially pressing keys of the key boards, code signals standing for words were sent to the computer, which contained the vocabulary, the grammar, our system for checking the correctness of sentences, and the rules for responding to some two dozen requests the chimpanzee Lana was to formulate in Yerkish word symbols. Pisani achieved the miracle of fitting all this into the minute memory of a PDP computer (Glasersfeld, 1977; Pisani, 1977).[3]

For six years we worked with the primatologists and the technicians of the Yerkes Center and the accomplishments of the chimpanzee Lana caught the attention of the press and TV. It was indeed absorbing and enjoyable work. But then there came a point when we resigned from the project, because irreconcilable differences had arisen regarding the direction of the research which remained firmly embedded in the behaviourist tradition. Nevertheless, I have no doubt that I owe whatever reputation I have gained in the field of psychology to the remarkable talents of Lana.

My background in conceptual analysis, however, bore another fruit. Michael Tomasello, one of the students whose master's thesis and dissertation I had the pleasure of directing, undertook the gigantic task of recording and analysing, together with his wife, all the first linguistic manifestations of their daughter during the second year of her life. To my knowledge, it constitutes the only complete database of early language acquisition, and it has proven a gold mine for the development and testing of theories in that area. It was an invaluable opportunity to see just how useful Ceccato's approach to the construction of concepts that we had further developed and expanded in our computer procedures would be in the analysis of children's conceptual development. Tomasello's recent book on a central topic of language acquisition, *First Verbs: A case study of early grammatical development* (1992), may well turn out to be a landmark.

Discovering Piaget

To the late Charles Smock, whom I remember with much affection, I owe my introduction to the work of Jean Piaget. It seems ironic that I should have had to come to America and to lose my research job in order to be introduced to the author who was to influence my later thinking more than any other. One evening, in one of the many long talks I had with Charles about language and epistemology, he said: 'It's funny, quite a few of the things you say I have heard from Piaget.' So I began reading Piaget — and since Charles had a large collection of texts he had acquired in Geneva, I read Piaget in French.

In the years that followed, when I came to teach courses on Piaget to students who could read only English, I realized how difficult, if not impossible, it is to understand the Piagetian orientation from translations. With very few exceptions (e.g., Wolfe Mays or Eleanor Duckworth) the translators seem to have a naive (i.e., naive realist) theory of knowledge and unconsciously bend what they read in Piaget's original texts to fit their own view of the world. As they cannot do this with everything, their translations often convey ideas that are incompatible with Piaget's theory or are downright incomprehensible.[4]

One example that might be of interest to mathematics educators is the translation of Piaget's book *The Child's Conception of Number* (1952a). There is first of all the inexplicable omission, on the cover and elsewhere, of the second author's name, Professor Alina Szeminska (Warsaw University), of whom Piaget says in his foreword that it was *her* talent that made possible the development of the particular methods of analysis used in the work. Then there is the recurrent mistranslation of individual terms, e.g., 'graduation' instead of gradation, 'equating' instead of equalizing. Most serious, given the topic of number, is the translators' indiscriminate use of the word 'set' for the French expressions for collection, quantity, row, sequence, series, and others. Small wonder that English-speaking mathematicians who read the translation thought: Who is this clown who writes about number when he doesn't even know what a set is!

The reader's understanding is further sabotaged by the translators' frequent omission of explanatory phrases and sometimes whole paragraphs. The unacceptable translation of this and other volumes provided part of my motivation for trying to present Piaget's thought to English students in a less distorted fashion. The correction of mistranslations, however, was not my primary goal. Having to teach Piaget from English textbooks, my main objective was to correct some of the basic misunderstandings concerning the nature of the constructivism that forms the backbone of his 'genetic epistemology'.

From Mental Operations to the Construction of Reality

Piaget was not the first to suggest that we construct our concepts and our picture of the world we live in, but no thinker before him had taken a developmental approach. To someone who is driven to ask about the source and validity of knowledge by circumstances of experience (in my case, the plurality of languages), it seems obvious that the best and perhaps only way to find out how knowledge is built up, would be to investigate how children do it. For traditional philosophers, of course, this would be committing an unforgivable sin, because to justify knowledge through its development rather than by a timeless logic, is what they call a 'genetic fallacy'. But Piaget was not a traditional philosopher.

In his *'La Construction du Réel chezl 'enfant* (Construction of Reality in the Child) (1937/1954) he presented a model of how a basic scaffolding — the conceptual structure of objects, space, time, and causality — can be built up. It serves as the framework within which a coherent experiential reality can be constructed. But this construction is not free, it is inevitably constrained and limited by the concepts that constitute the scaffolding. This is one of many points of overlap with the *A Theory of Personality: Psychology of Personal Constructs* of George Kelly (1963), who expressed the idea in the most general way:

To the living creature, then, the universe is real, but it is not inexorable unless he chooses to construe it that way. (Kelly, 1963, p.8)

Piaget himself explained what he intended by his genetic epistemology:

So, in sum, genetic epistemology deals with both the formation and the meaning of knowledge. We can formulate our problem in the following terms: by what means does the human mind go from a state of less sufficient knowledge to a state of higher knowledge? The decision of what is lower or less adequate knowledge, and what is higher knowledge, has of course formal and normative aspects. It is not up to psychologists to determine whether or not a certain state of knowledge is superior to another state. That decision is one for logicians or for specialists within a given realm of science. (Piaget, 1970a, pp.11–12)[5]

Having been prepared by Vico, Berkeley, Wittgenstein, and Ceccato, I read the quoted passage as a natural extension of a statement Piaget has repeated in many places, namely that knowledge is *not* a picture of the real world.

When he confronts 'less sufficient' or 'less adequate' with 'higher' knowledge, he is in fact saying that the *meaning* or value of knowledge lies in its function; and when he says that the adequacy of knowledge must be evaluated by logicians or scientists, he is simply explaining that it must be tested for logical consistency (non-contradictoriness) and for experiential validity (e.g., in experiments).

People who are tethered to traditional epistemology seem to be able to read this passage without being shaken in their belief that the better knowledge gets, the better it must represent an ontological reality. Hence, in their writings about Piaget, be they followers or critics, they disregard that, having started as a biologist, he saw cognition as an instrument of adaptation, as a tool for fitting ourselves into the world of our experience.

Because 'adaptation' and 'adapted' are frequently misunderstood (see Chapter 2) and 'adequate' tends to be interpreted as utilitarian, I adopted the biologists' term 'viability'. Actions, concepts, and conceptual operations are viable if they fit the purposive or descriptive contexts in which we use them. Thus, in the constructivist way of thinking, the concept of viability in the domain of experience, takes the place of the traditional philosopher's concept of Truth, that was to indicate a 'correct' representation of reality. This substitution, of course, does not affect the everyday concept of truth, which entails the faithful repetition or description of a prior *experience*.

For believers in representation, the radical change of the concept of knowledge and its relation to reality, is a tremendous shock. They immediately assume that giving up the representational view is tantamount to denying reality, which would indeed be a foolish thing to do. The world of our

Concept of viability takes the place of Truth

experience, after all, is hardly ever quite as we would like it to be. But this does not preclude that we ourselves have constructed our knowledge of it.

Radical constructivism, as I said at the beginning, is a way of thinking about knowledge and the act of knowing.

Because of its breach with the philosophical tradition it was (and still is in many quarters) quite unpopular. When I first submitted papers to journals, I received innumerable rejection slips from editors, one of whom stated his objection with endearing clarity: 'Your paper would be unsuitable for our readers.'

Having come from Europe and without background in psychology, it took me some time to discover what the problem was. In 1967 and for the decade that followed, the intellectual climate that pervaded departments of psychology and linguistics in the United States was largely dominated by behaviourism. As late as 1977, Skinner reiterated: 'The variables of which human behaviour is a function lie in the environment' (Skinner, 1977, p.1). If one believed in that kind of determinism, there could be no room for theories of mental construction. However, the belief in environmental determinism would make sense only if one had access to an objective environment, so that one could show that a particular piece of that environment causes a particular behaviour. But what a scientist — or indeed any reflective person — categorizes as his or her environment and then causally relates to observed behaviour, is always a part of that observer's domain of experience and not an independent external world.

A Decisive Friendship

The constructivist way of thinking on which I had been launched by Ceccato and Piaget obviously had no chance of making inroads upon the established dogma of the disciplines to which, I felt, it might have something to say. If Piaget himself had not succeeded in being taken seriously as a philosopher, it was clear that an obscure outsider could get nowhere. In linguistics, the work of Noam Chomsky had brilliantly turned the tables on the behaviourists. In doing so, however, he had posited the fundamental elements of language as innate, and this assumption closed the door to the constructivist approach. Psychology still bracketed the mind and proudly declared itself 'the science of behaviour'. In their textbooks, students were warned against the futility of philosophizing. Empiricism (which was usually understood as realism) was the password — and the one thing I did not have was empirical data to show the usefulness of the constructivist approach.

It was a great stroke of fortune that Charles Smock brought me together with Leslie Steffe. Once more a meeting profoundly influenced my life and work. Steffe ran a Piagetian research project in the Department of Mathematics Education at the University of Georgia. When we started talking about cognitive development and conceptual analysis, we at once discovered a vast area of agreement. For me, this was not only an encouragement but it came to form the basis of a collaboration that has been a truly wonderful experience ever since.

Whenever I think of this collaboration, I remember the passage in which the Nobel laureate Sir Peter Medawar demolished the popular image of the scientist 'as a regular, straightforward, plain-thinking man of facts and calculations'. Rather, he wrote:

> A scientist commands a dozen stratagems of inquiry in his approximation to the truth, and of course he has his way of going about things and more or less of the quality often described as 'professionalism' — an address that includes an ability to get on with things, abetted by a sanguine expectation of success and that ability to *imagine* what the truth might be which Shelley believed to be cognate with a poet's imagination. (Medawar, 1984, pp.17–18)

Substitute 'viable explanation' for the word 'truth', and you have what to me is a perfect portrait of Les Steffe. In the course of the innumerable fierce arguments we had throughout the years, we gradually expanded and refined each other's thinking, struggling at times for days to formulate our ideas so that they might become acceptable to both of us. There was also John Richards, a trained philosopher, who for long stretches participated in the battle; and Paul Cobb and Patrick Thompson, both graduate students of Steffe's, lived through the gruelling months of discussion that helped to forge a plausible model of what children might be doing on their way towards a concept of number and the basic operations of arithmetic (Steffe *et al.*, Steffe, Richards and Glasersfeld, 1978; Steffe, Thompson and Richards, 1982; Steffe, Glasersfeld, Richards, and Cobb, 1983).

Though I knew nothing about research in education and remembered little from the few semesters I had studied mathematics, there had always been in the back of my mind the idea that conceptual analysis would sooner or later have to deal with mathematical concepts. For a constructivist it was obviously impossible to think of numbers and geometrical forms as God-given. Nor could one accept the Platonic view of pure forms that float about as crystals in some mystical realm beyond experience. One would have to investigate their genesis as abstract entities in an experiential domain.

Mathematicians, from Euclid down to our time, are extremely non-committal about the make-up of their basic concepts. Numbers are the raw material of their abstract edifices, and much the way bricklayers do with bricks, they take them for granted.[6] Only they themselves could throw some light on how they arrived at their elementary concepts, but given their competence, the question obviously seemed trivial to them.

The philosophers I read, though some of them said quite clearly that number is 'a thing of the mind' (see Chapter 9), were no help either, because they did not explain how this mental entity could be produced. The one exception I found was Edmund Husserl, the founder of phenomenology, who proposed that the operation that forms discrete unitary objects in our perceptual field is essentially the same that underlies the concept of 'one' and, on a subsequent level of abstraction, enables us to comprise any collection of such ones in a discrete unitary entity that we call 'number' (Husserl, 1887, pp.157–68). This idea was certainly helpful and it fitted well into Piaget's theory of reflective abstraction. What was needed to substantiate it, was a domain of experience where such abstractions were likely to be made. Observing children was the answer.

Teaching Experiments

When I started working with Leslie Steffe, he was already far into the development of a method he called 'teaching experiment'. It was a hybrid of Piaget's 'clinical method' of interviewing children and educational research. Its goal was to establish a viable model of their constructive activities in the context of arithmetic. Steffe's approach differed in that its purpose was to create situations that would allow the investigator to observe children at work and make inferences as to how they build up specific mathematical concepts. If these concepts were abstractions from reflection rather than from sensorimotor experiences, it was clear from the outset that whatever inferences could be made about them would contain an element of conjecture. In time, however, the resulting hypothetical model achieves a high degree of plausibility and predictive usefulness.

> As with all general theoretical constructs, it is difficult to apply them to specific situations, when the cognizing subject is not ourselves but a 'subject' we are observing. In practice there may be observable behavioral indications, on the basis of which levels of abstraction can be determined, but making that determination is not simple. One might say that assuming something as 'given' or not is exclusively the subject's business. Hence, at best an observer can make educated guesses, taking into account — as does any experienced diagnostician — several indications collected over an extended period of observation. (Steffe and Glasersfeld, 1988, pp.18–19)

In the teaching experiments, there is no teaching in the conventional sense, and no curriculum. Yet many of the situations presented to the children contained arithmetical problems they might have encountered in school. What mattered, however, was not the solution but the children's untutored individual approach or, as Steffe says, the children's mathematics. The experiment proceeds, not along a fixed, preconceived plan, but the investigator has to invent it step by step according to what the child says or does. It is videotaped, and the real work takes place when the members of the team review the tapes and discuss them until they can agree on an interpretation.

It is easy to understand that reviewers of research proposals tend to be turned off when they are told that the methodology of the proposed research is the teaching experiment, but what the experiments will consist of cannot be foreseen, because it depends on the reactions of the subjects. We presume it was the quality of the publications emanating from Steffe's group (e.g., Steffe, Glasersfeld, Richards and Cobb, 1983; Steffe and Cobb, 1988) that has assured continued funding.

In the early 1970s, Piaget once more became fashionable in the United States, and this time the focus was on his constructivism rather than on the 'stage theory' that had previously been emphasized. As a result, a great many authors began to profess a constructivist orientation, though they seemed unaware of the principles of Piaget's epistemological position. Especially researchers in mathematics education assimilated the notion that children gradually build up their cognitive structures (hardly a novel idea), but they disregarded the fact that Piaget had changed the concept of knowledge. Consequently, when I was teaching genetic epistemology, I wanted to distinguish my approach from what students might be reading elsewhere about versions of constructivism that seemed trivial. I called the model I had been working on 'radical' and laid out its two basic principles:

- knowledge is not passively received but built up by the cognizing subject;
- the function of cognition is adaptive and serves the organization of the experiential world, not the discovery of ontological reality.[7]

The Spreading of Constructivist Ideas

In January 1978, Heinz von Foerster and Francisco Varela organized a conference in San Francisco that bore the title 'The construction of realities'. It was a closed symposium that brought together some thirty authors and scientists from a variety of disciplines who had in some way documented their belief that knowledge could not be found or discovered ready-made but had to be constructed.

It was a remarkable experience to learn that there were accomplished and widely respected thinkers in biology, sociology, political science, logic, linguistics, anthropology, and psychotherapy, who had in individually quite different ways come to the conclusion that the traditional epistemology could no longer be maintained. But, as so often in meetings of highly original minds, most of the time was spent on arguing about relatively small individual discrepancies, and very little on trying to formulate basic constructivist principles on which, it seemed, most if not all could have agreed.

For me, nevertheless, it was a most encouraging event. It was my only meeting with Gregory Bateson, and to listen to his comments and his wonderfully gentle way of pointing out a contradiction in a speaker's presentation was as important a lesson as the many insights he had provided in his writings.

It was also at this symposium that I met Paul Watzlawick, whose charming little book *How Real is Real?* (1977) I had shortly before used in a course of mine. Apart from being of Austrian origin, we had another thing in common: we had both lived and were still living in several languages. In many ways this liberation from a single mother tongue facilitates an immediate understanding of certain aspects of constructivism that take hard work and reasoning in all whose world view is constrained by a single language. Paul Watzlawick then invited me to write the introductory essay in his *The invented reality* (German edition, 1981; English, 1984), a book that has done more than any other to spread constructivist ideas.

Apart from my immeasurable debt to Leslie Steffe, I owe a great deal to the graduate students who worked with him and to those who took my courses in the Psychology Department. They were a motley lot, because many of them did not come from my department but from philosophy, linguistics, and mathematics education. Compared to one's peers — colleagues on the faculty and researchers who listen to presentations at professional meetings — graduate students are less inhibited. They tend to ask questions when what is said does not make sense to them. Thus they often put their finger on unconscious, unwarranted jumps in the development of ideas and inconsistencies in the presentation. No doubt this is of help to researchers in all disciplines. In the area of constructivism it is downright invaluable for a quite specific reason. I have often said that adopting the constructivist orientation requires the modification of almost all one has thought before. This is laborious and difficult to carry through. We are usually quite unaware of many habitual patterns of thinking in our minds. And there is another obstacle: the language in which our thoughts have to be formulated, be it English, Italian, or any other natural language, has been shaped by the naive realism inherent in the business of practical living and by a few prophets who were convinced to have access to an absolute reality.

For the very reason that radical constructivism entails a *radical* rebuilding of the concepts of knowledge, truth, communication, and understanding, it cannot be assimilated to any traditional epistemology. Above all, it seems enormously difficult to appreciate that it is not an orientation that claims to reveal an ultimate picture of the world. It claims to be no more than a coherent way of thinking that helps to deal with the fundamentally inexplicable world of our experience and, most important perhaps, places the responsibility for actions and thoughts where it belongs: on the individual thinker.

Retirement and a New Beginning

At the 1975 meeting of the Jean Piaget Society in Philadelphia, I for the first time presented the radical interpretation of genetic epistemology to a larger public. As it was a plenary session, there was not much discussion. But it had two important consequences for me. Hermine Sinclair, a long-time collaborator of Piaget, vigorously encouraged me to go on with my work, and it is to her that I owe my first invitation to Geneva. The presentation also led to a long talk with Jack Lochhead who was in the process of starting a Piagetian research group on cognition in the Physics Department of the University of Massachusetts. In the years that followed he invited me several times to give workshops on conceptual analysis, because he and his colleagues were trying to develop a more effective way of teaching physics and the mathematics it requires. Like other educational researchers, they had noticed that many students, were quite able to learn the necessary formulas and apply them to the limited range of textbook and test situations, but when faced with novel problems, they fell short and showed that they were far from having understood the relevant concepts and conceptual relations that constitute the actual framework of physics.

When I retired from the University of Georgia in 1987, Jack Lochhead called and said, why don't you come and work with us? By then his group had been established as an independent institute of the University at Amherst. The research in physics education, the modest title 'Scientific Reasoning Research Institute', the promise of snow and skiing in Massachusetts, and the fact that Charlotte's children would be much closer there, proved irresistible.

Work in the Institute soon made clear to me that teaching physics was not quite the same as teaching arithmetic in the elementary grades. Although the fundamental concepts in both areas are abstract constructs, their use is markedly different. In mathematics, concepts can be combined and related in all the ways the mathematician deems legitimate within the rules he has accepted; and new abstractions from such compounds may yield new levels of operating. If some of the resulting abstract structures are found to be applicable to worldly problems, this may be gratifying to their inventor but it is irrelevant to the pursuit of mathematics. In physics, however, the process of abstraction is doubly constrained. Not only must it comply with logic and be conceptually coherent, but its results must also withstand experimental tests, that is to say, they must fit experiential situations. In short, mathematics is self-sufficient and its goals lie within its own domain. Physics, in contrast, has an instrumental component, in that it has to provide theoretical models that help to organize our experiential world.

Support from Physics and Philosophy of Science

Although the problems that made up the daily work at the Institute were usually concerned with high-school physics and did not involve relativity or quantum theory, I had the opportunity to sit in on colloquia and meetings where such topics were discussed. Much of what I heard there seemed to confirm that it had not been misinterpretation, when years before I had picked a quotation here and there from the writings of the great physicists in our century. They all have at some point made statements that indicate they were aware of inventing or constructing theory rather than having it forced upon them by a collection of data. Werner Heisenberg, for instance, wrote:

> In the natural sciences, then, the object of research is no longer nature as such, but a nature confronted by human questions, and in this sense, here too, man encounters himself. (Heisenberg, 1955, p.18)

For the constructivist, such statements are a welcome corroboration, even if the same physicists, in their day-to-day work adopted a far more realist stance.[8] It is not at all surprising that a problem-solver takes *for real* the problematic experiential situation with which he is struggling. His task is technical, it lies within a specific, circumscribed area of experience, and its solution would not be advanced by epistemological considerations.[9] Only when he has solved it, may he adopt a philosophical attitude and conclude that his solution is an instrument for the organization and 'explanation' of experience rather than a representation of reality.

From the outset, it was clear that, in developing the constructivist approach to knowledge, the philosophy of science was an area that could not be skirted. Among my eclectic readings, one author in particular provided an irresistible challenge.

Karl Popper (1968) gave an excellent description of the instrumentalist view of science and then attempted to show that it was logically flawed. An attentive reader, however, will notice that his refutation is ultimately based on nothing but his metaphysical belief that scientific theories *can* approximate a rightness ('truth') that lies beyond the level of instrumental viability in given situations. The cornerstone of this belief are his notions of falsifiability and crucial tests. He claims that the fact that 'Newton's theory was falsified by crucial experiments which failed to falsify Einstein's' means more than that it broke down under certain experiential circumstances. Instrumentalism, he says, has nothing equivalent to such tests.

> An instrument may break down, to be sure, or it may become out-moded. But it hardly makes sense to say that we submit an instrument to the severest tests we can design in order to reject it if it does not stand up to them: every air frame, for example, can be 'tested to destruction', but this severe test is undertaken not in order to reject every frame when it is destroyed but to obtain information about the frame (i.e. to test a theory about it), so that it may be used *within the limits of its applicability* (or safety).

> For instrumental purposes of practical application a theory may continue to be used *even after its refutation*, within the limits of its applicability: an astronomer who believes that Newton's theory has

turned out to be false will not hesitate to apply its formalism within the limits of its applicability . . .

Instruments, even theories *in so far as they are instruments*, cannot be refuted. The instrumentalist interpretation will therefore be unable to account for real tests, which are attempted refutations, and will not get beyond the assertion that *different theories have different ranges of application*. But then it cannot possibly account for scientific progress. (Popper, 1968, pp.112–13)

On the strength of this, Popper concludes that instrumentalism is an 'obscurantist philosophy' (p.113).

For me, this passage was truly illuminating. Clearly, Popper had fully understood the thrust of instrumentalism. His example of the astronomer was an accurate prediction of how the scientists and engineers of NASA went about their business when they sent a man to the moon: they calculated everything according to Newtonian formulas, because this was far simpler and less time-consuming than using Einstein's, although of course they all knew that Newton's theory of the planetary system was no longer considered to be true.

Yet, because Popper is wedded to the notion of scientific progress, he is compelled to use the term 'refutation' in two different senses. On the one hand, it is the generation of circumstances in which a theory, used as an instrument, does not procure the desired result (or, in my way of speaking, is not viable). On the other hand, it is a crucial experiment that shows the theory to be *false* (where 'false' is interpreted as the opposite of 'true').

This is where a metaphysical belief enters the game. Indeed, Popper himself seems to have been aware that he was fudging. At the end of the chapter that is to justify his belief in the 'growth of knowledge', he writes, albeit in a footnote:

I admit that there may be a whiff of verificationism here; but this seems to me a case where we have to put up with it, *if we do not want a whiff of some form of instrumentalism that takes theories to be mere instruments of exploration*. (ibid., p.248, my emphasis)

That is the difference. Radical constructivism is uninhibitedly instrumentalist. It replaces the notion of 'truth' (as true representation of an independent reality) with the notion of 'viability' within the subjects' experiential world. Consequently it refuses all metaphysical commitments and claims to be no more than one possible model of thinking about the only world we can come to know, the world we construct as living subjects. Because this is a difficult and shocking change of attitude when one first comes to it, I want to reiterate once more that it would be misguided to ask whether radical constructivism is true or false, for it is intended, not as a metaphysical conjecture, but as a conceptual tool whose value can be gauged only by using it.

Notes

1 They were Giuseppe Vacccarino (logician), Ferruccio Rossi-Landi (linguist), Enzo Morpurgo (psychologist), Vittorio Somenzi (physicist), Enrico Maretti (engineer), Enrico Albani (computer scientist). Their subsequent publications that are relevant to my topic are listed among the references.

2 The journal lived for fifteen years and has recently been revived under the name *Methodologia* by Ceccato's student, Felice Accame.

3 It may seem incredible today, but all the computer memory that was available for the language system was 4 kilobytes.

4 I myself, indeed, have sinned in this respect, because for quite some time I translated Piaget's French *intelligence* as 'intelligence', forgetting that in many contexts it has to be read as 'mind' because that noun is not available in French.

5 Unless otherwise indicated, the translations of quotations from French, Italian, and German texts are mine.

6 The outstanding exception, of course, is the 'intuitionist' mathematician, L.E.J. Brouwer, but I did not become aware of his relevant paper (Brouwer, 1949) until after I had published my 'attentional model' (Glasersfeld, 1981a).

7 Although I had used this definition in lectures and talks, it did not appear in print until 1989, in my piece on constructivism in the *International Encyclopaedia of Education* (1989a), Supplement 1, p.162.

8 I have used similar quotations from Helmholtz, Mach, Einstein, and Bridgman in my papers, and others can be found in the philosophical writings of Bohr, Dirac, Born, and Schrödinger.

9 Note that I am using 'technical' to refer to the technique or method of science, not to machines and technology.

Chapter 2

Unpopular Philosophical Ideas: A History in Quotations

In the first chapter I recounted how biographical circumstances — my up-bringing, living in certain places, meeting a few exceptional people, and eclectic reading — led me to an unconventional way of thinking. Yet, there is nothing new about the ideas that make up radical constructivism. The only novelty may be the way they have been pulled together and separated from metaphysical embroidery.

I agree with Bertrand Russell's definition:

> Metaphysics, or the attempt to conceive the world as a whole by means of thought, has been developed, from the first, by the union and conflict of two very different human impulses, the one urging men towards mysticism, the other urging them towards science . . . the greatest men who have been philosophers have felt the need both of science and of mysticism: the attempt to harmonise the two was what made their life, and what always must, for all its arduous uncertainty, make philosophy, to some minds, a greater thing than either science or religion. (Russell, 1917/1986, p.20)

However, I do not agree that to attempt such a union would be a *rational* undertaking. For me, whatever is truly mystical eludes the grasp of reason. This is neither a denial nor a judgment of value, it merely expresses the conviction that the mystical is a closed domain of wisdom that withers under the cutting tools of reason. The purpose of reason is analysis. Whatever reason wants to deal with must be describable in terms of specific differences and therefore has to be articulated into entities and relations. The mystical treats the world as a whole that requires no differentiation from any background. When it speaks of parts, they are metaphors intended to generate empathy with the ultimate oneness.

Radical constructivism is intended as a model of rational knowing, not as a metaphysics that attempts to describe a real world. I believe that its effort to delimit the purview of reason is one of its virtues, precisely because this limitation accentuates the need to contemplate the realm of the mystic's wisdom.

The history I shall present, therefore, is an attempt to justify this separation.

It will not be complete in any sense. In part, this is because I am citing only those authors who seemed indispensable to me — and others could have been included to strengthen particular points; in part, it will be incomplete because there are authors (e.g., Collingwood, Dewey, and Peirce) who undoubtedly said related things, but whom I do not know well enough to discuss.

Objectivity Put in Question

Once one steps out of the philosophical tradition and questions the illusory goal of attaining true representations of a real world, quite a few thinkers can be found who have taken a step in this direction. Most of them, however, ran into a serious problem. By renouncing the quest for certain knowledge about reality, they had deprived themselves of the very argument philosophers use to distinguish knowledge from mere opinion or belief. Consequently, these wayward thinkers were for the most disregarded in the history of philosophy and, at best, left by the wayside as oddities. The traditional way of thinking was (and still is) far too strong to be shaken by a critique that offers no immediate replacement.

In the last hundred years, the situation has begun to change. During the nineteenth century, science was regarded as a sophisticated extension of common sense that had gradually unveiled the mysteries of the real world. The successes of technology seemed an unquestionable confirmation of realist epistemology. Then, however, came spectacular scientific developments — especially in theoretical physics — that engendered internal doubts in the representational character of scientific explanations. Could science unveil the character of the world *as it is*? The passage from Heisenberg I quoted in the preceding chapter suggests that the scientist cannot escape the human ways of seeing and thinking. Objectivity, thus, became doubtful. Towards the end of his life, Jacob Bronowski described the changed situation:

> There is no permanence to scientific concepts because they are only our interpretations of natural phenomena . . . We merely make a temporary invention which covers that part of the world accessible to us at the moment. (Bronowski, 1978, p.96)

Today the very philosophy of science teems with ideas that subvert the millenary tradition of realism and its goal of objective knowledge. In the face of this turmoil, it may be both legitimate and appropriate to review the history of epistemological dissent.

To me, such a review is of particular interest. Not because I hope to find many pioneers of constructivism, but rather because the record of thinkers who went against the established view, confirms the need for a radically different approach to the problems of knowing.

The Pre-Socratics

From the beginnings of western philosophy, there was one group of dissidents that could not be ignored. Their arguments were logically incontrovertible. They were the 'sceptics', and their first school was founded by Pyrrhon towards the end of the fourth century BC. Its teachings were assembled and annotated more than five hundred years later by Sextus Empiricus.

The sceptics collected innumerable examples to demonstrate the unreliability of the human senses, showing that perceptions and the judgments based on them were influenced by context and human attitudes and therefore could not be trusted to provide a true picture of the real world. If, for instance, you move your hand from a basin of cold water into one that is tepid, the second feels hot; if you begin with hot water, the tepid feels cold; the *true* temperature of the water, therefore, cannot be determined because our judgment depends on the experiential context.

The belief that true knowledge of the real world could never be attained, had actually been expressed in the most succinct way by Xenophanes, who lived some two hundred years earlier than Pyrrhon:

> Certain truth [about God or the world] has not and cannot be attained by any man; for even if he should fully succeed in saying what is true, he himself could not know that it was so. (Xenophanes, Fragment 34)[1]

In the course of the centuries, the gist of this statement has cropped up in many forms. It has invariably been attacked by means of a spurious argument. If you deny that there can be certain knowledge, it is claimed, you cannot be certain of this denial either. The argument is spurious, because it confounds the domain of logic and mathematics with the domain of certain knowledge about the world. Xenophanes would have had no qualms about admitting the certainty of $2 + 2 = 4$, because once agreement has been reached about the rules that govern counting and on a fixed sequence of number words, everyone is obliged to arrive at 'four' whenever two pairs of items are counted in sequence. This certainty reveals something about the numbering system invented and agreed on, but it reveals nothing about God's flair for mathematics or a world of numbers supposed to be independent of the counters.

Indeed, the reasoning that underlies Xenophanes' insight involves the logic of thinking, not the particulars of experience. To claim true knowledge of the world, you would have to be certain that the picture you compose on the basis of your perceptions and conceptions is in every respect a true representation of the world as it *really* is. But in order to be certain that it is a good match, you should be able to compare the representation to what it is supposed to represent. This, however, you cannot do, because you cannot step out of your human ways of perceiving and conceiving.

About a hundred years later, Protagoras, the first of the Sophists in the fifth century BC, formulated the famous phrase:

Man is the measure of all things. (Protagoras in Guthrie, 1971, p.171)

Today, we might say: a human being's view of the world is necessarily a *human* view. Unless you claim some form of direct mystical revelation, whatever you call knowledge — your ideas or concepts, the relations that connect them, your images of yourself and the world — will be human, because the way you have produced them was yours, and you, whether you like it or not, are bound by the human ways.

All great philosophers of the western world have admitted the logical irrefutability of this argument. Nevertheless, they have struggled to find a way around it. In one form or another, explicitly or surreptitiously under the guise of metaphysics, they resorted to mysticism or religious revelation.

Plato was apparently aware of the paradoxical character of the concept of knowledge, and he tried to resolve it with the metaphor of a line divided in four parts (*The Republic*, 509d–517b). The first two segments represent the world of the senses: shadowy images of imagination and conjecture, and the forms of things we derive from perception. They are not real things, and he illustrated this by the famous parable of the cave. In this domain there is no certain knowledge but only 'opinion' (*doxa*). The third section holds the understanding of products of thought (*episteme*), such as mathematics. The fourth belongs to the eternal ideas of beauty, justice, and goodness, which are every human's heritage since God created the universe, and it is here that true wisdom may be achieved. The metaphor of the line was to suggest the possibility of development, as though one could escape from the shadows of the cave and come to see the divine Truth by the power of human reason.

Theological Insights

That there is no logical justification for such a belief in the powers of reason, was seen not only by the sceptics, but early on also by men of the Christian faith. In Byzantium, in the third century AD, a school of theology arose that later became known as 'apophatic' or 'negative' theology. It formulated a powerful principle:

> The absolute transcendence of God excludes any possibility of iden-
> tifying Him with any human concept . . . for no human word or
> thought is capable of comprehending what God is. (Meyendorff, 1974,
> p.11)

To put it simply: if God is omnipotent, omniscient, and present every-where at the same time, then He is different from all the things we encounter in the world we live in; and since our concepts are derived from living experi-ence, we cannot capture the character of the divine in those concepts.

Of course, the Byzantine theologians did not deny the possibility of revelation, but they made very clear that revelation was not to be confused with rational knowledge. The Church, that had always claimed to be the authorized interpreter of God's ways and God's will, did not appreciate this kind of theology, and declared it a heresy. But it survived, and echoes of it crop up in the writings of medieval mystics.[2]

Most extraordinary among them is the Irish scholar John Scottus Eriugena who was born early in the ninth century and spent much of his life in France. As a monk he was mostly concerned with theology and followed the 'negative' direction of the Byzantine fathers, but his theory of knowledge had a wider scope. He was interested in reason as such and in the kind of knowledge it could produce. Two quotations suffice to show how modern he was. The first is an uncanny anticipation of an insight that Kant (I believe, quite independently) formulated in the preface to his *Critique of Pure Reason* (1787):

> For just as the wise artist produces his art from himself in himself and foresees in it the things he is to make . . . so the intellect brought forth from itself and in itself its reason, in which it foreknows and causally pre-creates all things which it desires to make. (Eriugena, *Periphyseon*, Vol.2, 577a–b)[3]

The second passage from Eriugena foreshadows Descartes' famous '*cogito ergo sum*' (I think, therefore I am) but does not encourage the vain hope that the establishment of one's own existence could serve as a basis for the attainment of certain truths about the world:

> Man, like God, can know with absolute certainty that he is, but cannot circumscribe his nature so as to be able to say what he is. (Quoted in Kearney, 1985, p.97)

When the Byzantine thinkers asserted the impossibility of grasping the essential character of God by means of human concepts, they were doing theology. However, the argument that our concepts are formed on the basis of our experience and can therefore not be used to describe anything that lies outside our experiential field, applies not only to superhuman entities but also to any 'reality' we posit beyond the things that we experience. Eriugena then emphasized the fact that reason operates according to its own rules and cannot transcend them (see Kant, below).

Modern Science Widens The Rift

Thus, even before the year 1000 there was the suggestion that there are two different kinds of knowledge, but the division is not quite the division Plato had proposed. For him, sensory experience led to 'opinion', and reason to 'certain knowledge'. Now, we have the clear but fallible knowledge of experience and the eternal truths of mystical revelation.

The rift in the concept of knowledge was present but somewhat dormant through the Middle Ages (see McMullin, 1988, p.31). It became topical in the Renaissance, when Copernicus, Kepler, and Galilei proclaimed a model of the planetary system that was in direct contradiction to the teaching of the Church. Wise men like Osiander, the editor of Copernicus' posthumous work, and Cardinal Bellarmino, who wanted to help Galilei avoid a trial for heresy, tried to defuse the clash. The scientist, they said, was not committing heresy, as long as he used his theory to calculate predictions and to provide plausible models of phenomena.[4] The one thing he must *not* do, is claim that he is describing the reality of God's world, because God's world is the province of the Church and its dogma. It was the first clear assertion that the knowledge of science should be considered instrumental and fallible, whereas the mystical wisdom of revelation is unquestionable and an end in itself.

Though Galilei formally recanted, he was by no means willing to accept this partition. He did not want to give up the notion that scientific theory could describe the real world. There is a strange irony in this refusal.[5] Galilei is celebrated as the founder of modern science and his method has, indeed, been enormously successful. It can be characterized as a procedure that invents ideal entities whose behaviour is governed by ideal laws. These invented ideas are then used to explain the observed behaviour of experiential items, by introducing disturbances that prevent them from obeying the ideal laws. His law of 'free fall', for instance, demands that all bodies, regardless of their weight, accelerate at the same rate when they are falling. Galilei formulated the law, although there was no way he could have experienced exactly what the law said or demonstrated it experimentally — not even with the help of the leaning tower of Pisa. Nevertheless, if one added some factors representing considerations such as friction and air resistance to the ideal law, the calculations became eminently useful in a great many practical situations.

Torricelli, a famous student of Galilei's, expressed this very clearly:

> Whether the principles of the doctrine *de motu* be true or false is of very little importance to me. Because if they are not true, one should pretend that they are true, as we supposed them to be, and then consider as geometrical and not mixed [empirical], all the other speculations that we derived from these principles . . . If this is done, I say that there will follow everything Galileo and I have said. Then, if the balls of lead, of iron, or of stone do not behave according to our computation, too bad for them, we shall say that we were not talking of them. (Belloni, 1975, p.30)

The quoted passage would fit well into the discussions that today are enlivening the philosophy of science, discussions in which the objectivity of science and its theories is being put in question from a variety of positions. From the constructivist point of view, it is of course encouraging to find that, long before the invention of quantum theory, there were great physicists who did not quite believe, as Galilei did early on, that 'the book of nature is written in the language of mathematics', but were more inclined to think that mathematics was a rather neat human way of ordering and managing the human experience of nature.

The wide range of successful applications of the physicists' laws to puzzling problems in the field of experience, soon became the dominant factor that empowered the knowledge of science at the expense of the mystical. Philosophers, however, could not be satisfied with practical success. Bound as they were by a tradition of 2000 years, they were concerned with absolute Truth. Scientific theories, after all, were forever changing, and their confirmation by experiments could never allay the sceptics' argument against the reality of phenomena. Philosophers required eternal truths that would point to the solution of all problems, including those posed by science. But they were not prepared to admit that a religious or mystical faith was the way to grasp those truths. They did not want to do without God, but He had to operate within human reason.

A Failure and an Achievement of Descartes

Descartes, who was profoundly disturbed by the fact that some of his contemporaries were applying the rediscovered teachings of Pyrrhon's school to religious beliefs, decided to pursue the quest for certain knowledge in an uncompromising way. By subjecting all ideas to doubt, he hoped to isolate those that could not be doubted. He found only one: he could not question that it was he who was thinking the doubts. But when he tried to use this certainty to build up other ideas that were indubitable, he failed and had to resort to an act of faith. 'Since God is no deceiver', he said, 'the faculty of knowledge that He has given us cannot be fallacious' (Popkin, 1979, p.177).

Instead of demolishing scepticism, Descartes' method of doubt had enhanced it. But he achieved a number of other things that show his genius. One of them was the invention of analytical geometry, the ingenious way of translating geometry into algebra. I was told in high school how he came to make this invention.

The story is apocryphal but of immediate appeal to constructivists. When he was 23 years old, Descartes joined an army and was moved to southern Germany. There was no war at that time and he was billeted in a peasant house. It was winter and he spent most of his time not only indoors but, as he put it, 'in a stove'. This sounds odd, but if one knows peasant houses in that region, there is no mystery. A corner of the family room is usually taken up by a large tiled stove that has a wooden structure around it and a platform above, which is large enough to stretch out, a couple of feet or so below the ceiling. It is the warmest place in the house — and the flies know it. They use this part of the ceiling as a home base.

Lying on this platform, Descartes looked up at the ceiling and saw flies walk about. Having a mathematical bent, he asked himself how one could accurately describe their movements — and had a flash of genius. There were two lines, formed by the meeting of the walls and the ceiling, that met in a right angle in the corner of the room. The position of a fly could be described by projecting it on both lines and measuring the respective distances of the two projections from the corner. If the fly moved in a straight line and you applied the same procedure to the endpoint of its movement, you could express the fly's itinerary by means of the distances of the first and second projections on each of the axes.

It may well have been this experience which convinced Descartes that certain knowledge, if there was any, would have to spring from reason and its most perfect embodiment, mathematics. For the constructivist, the story — whether it is true or not — is a nice example of the principle that mathematical ideas can be abstracted from sensorimotor experiences.

Locke's Forgotten Reflection

(★ Maths in the everyday)

In the century after Descartes, there followed one after the other, the Englishman John Locke, the Irish bishop George Berkeley, and the Scot David Hume. Together, the three philosophers later became known as the 'British Empiricists'. This designation can be found in every introductory psychology text, and it has given rise to the notion, both among teachers and students, that the three men formed a harmonious team fighting for the same ideas. This is a poor description of the trio because each of the men disagreed with his predecessor as much as with other philosophers. More serious, however, is the spurious meaning that has been attributed to the term 'empiricism'.

Empiricists agree that knowledge springs from experience and that experience is its testing ground. But on the question of how experience is to be related to a real world beyond it, they may have widely different opinions. Today, however, one often reads the expression 'hard-nosed empiricist', and it is intended to convey that experimental evidence provides data that reflects the character or the state of an observer-independent real world. None of the three British empiricists was so naive a realist.

Locke, as far as I know, was the first to use the term 'reflection' with the meaning that is fundamental in cognitive constructivism since Piaget. His explanation is not the most transparent and requires some concentration:

> This source of ideas every man has wholly in him-self; and though it be not sense, as having nothing to do with external objects, yet it is very like it, and might properly enough be called *internal sense*. But as I call the other Sensation, so I call this REFLECTION, the ideas it affords being such only as the mind gets by reflecting on its own operations within itself. By reflection then, in the following part of this discourse, I would be understood to mean, that notice which the

mind takes of its own operations, and the manner of them, by reason whereof there come to be ideas of these operations in the understanding. (Locke, 1690, Book II, Chapter i, par.4)

Locke was well aware of the fact that Descartes (and Galilei) had discredited the reliability of the sensations of colour, taste, smell, etc., (secondary qualities) and he agreed that we merely

imagine that those ideas are the resemblances of something really existing in the objects themselves. (ibid., Chapter viii, par.25)

In contrast, the 'primary' qualities,

bulk, figure, number, situation, and motion or rest, . . . may be properly called real, original, or primary qualities; because they are in the things themselves, whether they are perceived or not: and upon their different modifications it is that the secondary qualities depend. (ibid., par.23)

He does not explain why he considers this less 'imagined' than the reality of the secondary qualities. Indeed, it is ironic that the father of empiricism here tacitly aligns himself with Plato's idealism and assumes that there are ideas that do not derive from experience.

The Exaggeration of the 'Blank Slate'

Much has been made of the slogan that the new-born child's mind is a 'blank slate' on which experience alone inscribes knowledge. Locke himself used expressions such as 'empty cabinet', 'white sheet', and 'waxed tablet', but in view of what he says about ideas springing from the mind's reflection upon its own operations, these metaphors are misleading (see Fraser, 1959; Vol.1; p.48). Relational notions, he says, are always the result of sensation *and* reflection. He explains this, for instance, in the case of cause and effect:

. . . finding that the substance, wood, which is a certain collection of simple ideas so called, by the application of fire, is turned into another substance, called ashes; i.e., another complex idea, consisting of a collection of simple ideas, quite different from that complex idea which we call wood; we consider fire, in relation to ashes, as cause, and the ashes as effect. So that whatever is considered by us to conduce or operate to the producing any particular simple idea, or collection of simple ideas, . . . which did not before exist, hath thereby in our minds the relation of a cause, and so is denominated by us. (ibid., Book II, Chapter xxvi, par.1)

This pattern of generation is repeated for several abstract concepts and it clearly involves sensation, as it starts with collections of 'simple ideas'. But it also involves the observer's reflective mind. In the quoted case it is the mind that considers the one collection to be what conduces or operates to produce the other and turns a simple sequence into the relation of cause and effect.

As far as 'understanding' is concerned, which is the main word in the title of Locke's work, the pattern of concept generation he illustrated in no way contradicts the image of the child's empty slate — it merely claims that the construction of knowledge cannot begin until there are some simple sensory ideas for the mind to operate. Only then can the mind reflect and abstract new complex ideas from its own operations. It fell to Berkeley, to point out Locke's inconsistency in the treatment of primary qualities.

The myth that Locke's empiricism held *all* knowledge to be derived directly from the senses was greatly reinforced by the misunderstanding of his use of the term 'experience'. For him, this included not only the acquisition of sense ideas but also their retention and subsequent elaboration by means of reflection and abstraction (see Fraser, 1959; p.49).

A Reinterpretation of Berkeley

George Berkeley, the second of the British empiricists, read Locke's *Essay concerning human understanding* at the very beginning of the eighteenth century, when he was studying at Trinity College in Dublin. He kept a notebook, and in it the barely 20-year old philosopher recorded early formulations of ideas he was then to develop and expound in his *Essay towards a new theory of vision* (1709) and the *Treatise concerning the principles of human knowledge* (1710).[6] There are also many entries that indicate agreements and disagreements with Locke. One of the major disagreements concerns the relation between the 'primary qualities' and real things.

I believe that Berkeley's objection to the notion that these qualities are less dependent on the observer and therefore 'truer' than the secondary ones, is here derived from a consideration that he never expressed more clearly in later writings.

> Extension, motion, time do each of them include the idea of succession, & so far forth they seem to be of mathematical consideration. Number consisting in succession & distinct perception wch also consists in succession, for things at once perceiv'd are jumbled & mixt together in the mind. Time and motion cannot be conceiv'd without succession, & extension . . . cannot be conceiv'd but as consisting of parts wch may be distinctly and successively perceiv'd. (Berkeley, 1706, par.460)

The expression 'mathematical consideration' becomes clear if one takes into account that the quoted paragraph 460 is the answer he provides to a question about extension which he asked himself in the much earlier paragraph 111, as a query immediately after:

> Number not in bodies, it being the creature of the mind depending entirely on its consideration & being more or less as the mind pleases. (ibid., par.110)

Berkeley was well aware that all mathematical thinking results from reflection and abstraction. When he realized that *succession* could not be a property of sensory objects but had to be abstracted by a subject's reflection upon its own experience, he called it a mathematical notion, even where it gave rise, not to numbers, but to concepts such as extension, motion, and time. The important point in this is the realization that the features that were considered primary (in the sense that they reflect properties of real objects) depend on concepts that are formed from a succession of at least two experiential frames and an act of relating them. The succession then merely provides the experiencing subject with an opportunity to establish a relation; it does not require it. Nor does the succession itself determine what kind of relation should be established.

It is often said that Berkeley demolished the objectivity of primary qualities by showing that the arguments Locke had used against the secondary qualities were equally effective when applied to the primary ones. I would not deny this. But his insight that the elementary conceptual relations of extension, motion, time, and causation are not supplied by the mere succession of experiences, seems a far more powerful argument to me. The realization that these basic relational building blocks have to be generated by the experiencing subject, wipes out the major rational grounds for the belief that human knowledge could represent a reality that is independent of human experience. For if extension, motion, time, and causation are dependent on the reflective activity of a subject, one cannot describe in human terms what 'reality' would be like *before* it is experienced.

Hume's Deconstruction of Conceptual Relations

The interpretation of succession is, indeed, one of the problems that led David Hume, the third British empiricist, to his most uncompromising scepticism. He made a valiant effort to unravel how we come to impose specific relations:

> Tho' it be too obvious to escape Observation, that different ideas are connected together; I do not find, that any Philosopher has attempted to enumerate or class all the Principles of Connexion; a Subject, however, that seems very worthy of Curiosity. To me, there appears to be only three Principles of Connexion among Ideas, viz. *Resemblance*,

Contiguity in Time or Place, and *Cause* or *Effect.* (Hume, 1742, Essay III)

He proceeds to give examples of how these 'connexions' are made and then, before concluding the Essay by once more listing the three 'principles', he says:

These loose Hints I have thrown together, in order to excite the Curiosity of Philosophers, and beget a Suspicion at least, . . . that many Operations of the human Mind depend on the Connexion or Association of Ideas, which is here explain'd. (ibid.)

Later in his work, when he discusses 'the communication of motion by impulse, as by the shock of two billiard balls' he states:

When, therefore, we say, that one Object is connected with another, we mean only, that they have acquir'd a Connexion in our thoughts, and give rise to this inference, by which they become Proofs of each other's Existence. (ibid., Essay vii, Part I)

It is crucial to remember that Locke and Hume were concerned with human *understanding*, Berkeley with human *knowledge*. All three focus primarily on how the rational mind acquires knowledge and how knowledge is constituted. When Hume, in the context of the quoted passage, speaks of 'existence', it is the existence that Berkeley has defined as perceivability in the domain of experience, and not ontological *being*. That this interpretation is justified, becomes clear when one considers a later passage in Hume, concerning the question whether the perceptions of the senses are produced by external objects that resemble them:

How shall this Question be determin'd? By experience surely; as all other Questions of a like Nature. But here Experience is, and must be entirely silent. The Mind has never any thing present to it but the perceptions, and cannot possibly reach any Experience of their Connexion with Objects. The Supposition of such a Connexion is, therefore, without any Foundation in Reasoning. (ibid., Essay xii, Part I)

After this, the belief that human knowledge ought to represent an absolute reality could no longer honestly be justified by reasoning about experience, but had to find support in the realm of metaphysics.[7] The realization that 'relating' is under all circumstances a conceptual act and therefore requires an active mind to conceive it, was no doubt one of the factors that prompted Kant to say that Hume had shaken him out of a 'dogmatic slumber' (Kant, 1783, p.260).

Bentham and Vico — Pioneers of Conceptual Analysis

In 1780, a year before Kant published his *Critique of Pure Reason*. Jeremy Bentham produced a 'preliminary' treatise on jurisprudence which contained first elements of his 'Theory of fictions'. The theory as a whole was pieced together from a variety of writings that appeared in the last two decades of the eighteenth century and until after his death in 1832 (see Ogden, 1959; p.XXff). To my knowledge, Bentham initiated the *operational* analysis of concepts and thus took the first step in a direction that would later help to resolve a major problem in Kant's philosophy: the assumption of a priori categories. For Bentham, too, the concept of relation was a trigger:

> No two entities of any kind can present themselves simultaneously to the mind — nor can so much as the same object present itself at different times — without presenting the idea of *Relation*.[8] For relation is a fictitious entity, which is produced and has place, as often as the mind, having perception of any object, obtains, at the same, or, at any immediately succeeding instant, perception of any other object, or even of that same object, if the perception be accompanied with the perception of its being the same: *Diversity* is, in the one case, the name of the relation. *Identity* in the other case. (Bentham, in Ogden, 1959; p.29)

Bentham's analyses were 100 years ahead of his time (Ogden, 1959; p.cli) and had to wait for Hans Vaihinger (see below) to be appreciated and further developed.

Far from Britain, there was another thinker in the first half of the eighteenth century whose work anticipated some of the most important ideas of constructivism. In 1710, the year of Berkeley's 'Treatise', Giambattista Vico published the Latin tract *De antiquissima Italorum sapientia* in Naples and opened a new perspective on epistemology. It had little resonance in Italy and, until recently, remained almost unknown in the English-speaking world. On the strength of his later work, Vico became known as a seminal thinker in the philosophy of history and sociology. His theory of knowledge pervaded all he ever wrote, but because he did not again devote a special text to it, it was mostly misinterpreted and treated as a marginal curiosity by his readers and commentators.

Vico, too, was perturbed by the inroads of scepticism on matters of religious faith and revelation, but he profoundly disagreed with Descartes. Instead of doubting everything in order to find certain truth, he wanted to separate the mystical from the rational once and for all. To do this, he first of all set out to specify what should be considered characteristic of the two domains. Second, he had the brilliant idea of examining the means by which each of them expressed its products. Language, thus, became a key factor in his theory.

He opens his treatise on 'The most ancient wisdom of the inhabitants of Italy' with some etymological observations. For the speakers of Latin, he says, the words *verum* (true) and *factum* (fact) were interchangeable,[9] *intellegere* meant quite the same as 'to know', and:

> *Ratio*, for them, meant the composition of arithmetical elements and, as such, the faculty that is proper to man and puts him ahead of the animals. (Vico, 1710, Ch.I, par.I,1)[10]

Reason, he said, produces knowledge by finding out how things are put together or made. It specifies what they consist of and how the components are related among each other. For God, who created the world, making and knowing are one and the same, and His knowledge is infinite.

> This is the norm to which all human truths should be compared; this is to say, among human cognitions those are true, whose elements are within ourselves and co-ordinated by ourselves and which, by means of postulates we continue to produce ad infinitum; and as we put together these elements, we become the makers of the truths that we know by composing them.[11] (Vico, 1710, Ch.I, par.III,2)

Human reason, therefore, can know only those things that are made of material to which it has access — which is the material of experience — and it is through the making that the knowledge of them arises. As far as I know, Vico was the first to state unequivocally that our rational knowledge is constructed by ourselves. But he was also a religious man and had a stake in metaphysics. Hence he wanted to account for mystical knowledge. He did this in a way that is both simple and effective, by dividing knowledge in two kinds: 'rational knowledge', which regards the world of everyday experience and science; and 'poetic wisdom', which regards all that lies beyond that tangible world.

The rational can be expressed in 'vulgar language', by which he meant words that designate experiential things and the relations abstracted from them. The poetic, in contrast, is expressed in metaphors that point beyond the rationally accessible. He explicitly says:

> . . . to give utterance to our understanding of spiritual things . . . we must seek aid from our imagination to explain them and, like painters, form human images of them. (Vico, 1744, par.402)

Vico went to great lengths to show that, at the beginnings of human culture, all abstract knowledge was couched in poetic metaphors, the language of fables.[12]

One of his examples is the one that Joyce picked when he wrote *Finnegans Wake*. Thunder, Vico remarked, was both terrifying and inexplicable. It came

37

from the sky, and its origin could not be investigated. For the primeval mind, therefore, a superhuman power, a god, had to reside in the sky (ibid. par.377). So the sky became the abode of superhuman powers and the origin of things and events that could not be accounted for by explanations gathered by induction from everyday experience.

Vico calls such imaginative inventions 'metaphorical' because they are based on analogies of which only one half is accessible. Even the earliest human beings could hear a thunder-like noise when they pushed a boulder out of their cave and it rolled down the mountainside; and they *imagined* that thunder must be caused in a similar way (ibid. par.444). The similarity, however, is merely an assumption projected beyond the realm of experience where it could be checked. It is not an hypothesis but a fable. When such fables are repeated from generation to generation and coordinated with one another, a mythology is created, its origin in poetic imagination is forgotten, and it is taken as knowledge that was abstracted by someone from actual experience.

Vico had many other seminal thoughts. He suggested, for instance, that geometry was constructed on the basic mental operation that generates a point, and that this was the same operation that generated a moment in time. He anticipated the Piagetian notions that development could always be described in stages and that the human mind assimilates the unknown to the concepts with which it is familiar. In my view, however, his most powerful contribution to the analysis of ideas was the clear distinction he provided between metaphors based on an analogy in experience and the poetic ones of the mystic (or metaphysician) who projects things and events by means of analogy into the unknowable.[13]

Kant's 'Transcendental Enterprise'

So far, I have used the term 'metaphysics' to refer to attempts to describe the world as a whole, including both the domain of human experience and whatever is supposed to lie beyond. In Kant's work, the term becomes more complex because he uses it in explicitly differentiated senses which he characterizes by means of two expressions that are easily confused: transcendental and transcendent. He calls his analytical/critical investigations of reason 'transcendental philosophy' and specifies that this deals with:

> understanding and reason itself as a system of concepts and principles that regard objects in general, *without the assumption of things that might be given [ontology]*. The second (the *transcendent*) regards Nature, i.e., the sum of given objects — *whether they be given to the senses or, if you will, to some other kind of intuition*. (Kant, 1787; p.873; my emphasis)

He repeats several times that all that belongs to this second, 'transcendent' part is 'speculative' and 'goes beyond the bounds of possible experience'. In my view, the second part is rationally unconvincing because it has to employ concepts and language that were derived from experience, and the use of such means 'beyond the bounds of experience' therefore implicitly involves the assumption that the range of their application stretches beyond the domain in which they were formed. I agree with Vico and claim that whatever is transcendent can be spoken of only in poetic metaphors and therefore belongs to the realm of the mystical.

Kant's 'transcendental philosophy', however, is a purely rational analysis of human understanding and provides a model that is in many ways fundamental to the constructivist orientation.

In the preface to his *Critique of Pure Reason*, Kant observed that, as far as he could see, all prior attempts to investigate the products of our *knowing*, i.e., our cognitions, did not progress with 'the sure tread of a science' (1787, p.vii).[14] One reason for this is that:

> Until now one assumed that all cognition had to conform to objects . . . Henceforth one might try to find out whether we do not get further . . . if we assume that the objects have to conform to our cognition. (Kant, 1787, p.xvi)

Of Galilei, Torricelli, and other scientists he says that they 'saw the light':

> They understood that reason can comprehend only what she herself has brought forth according to her design . . .

> . . . On the one hand, reason must proceed according to her principles, and only in accordance with them may appearances count as laws; and on the other hand, reason must approach Nature, not like a student who listens to whatever the teacher wants to say, but rather like an appointed judge who compels the witnesses to reply to the questions he puts to them. (Kant, 1787, p.xiii)

The first three lines of this quotation could be a summary of what Eriugena wrote a thousand years earlier. Then, it was intended to show that reason could not encroach on the mystic's wisdom. For Kant, it was the premise that led to his analysis of all rational knowing, summarized in his *The conflict of the faculties* (1798).

> The understanding is a wholly active power of the human being; all its ideas and concepts are but its creation, . . . External things are only occasions that cause the working of the understanding . . . the product of its action are ideas and concepts. Thus, the things to which these presentations (*Vorstellungen*)[15] and concepts refer cannot be what our mind presents to itself; because the mind can create only presentations

of its own objects and not of real things, that is, through these presentations and concepts, things cannot possibly be known as they might be in-themselves. (Kant, 1798, *Werke*, Vol.vii, p.71)

This leaves open the question why things should be considered 'external' at all, and not simply figments of human imagination. Kant's answer is extremely complicated and therefore open to a variety of readings that I consider to be misinterpretations.

The things our senses and our understanding present to us . . . are the product of the coming together of causal occasions and the effect of understanding. Nevertheless they are not [merely] appearance, for in our practical life we can consider them real things and objects of our presentations — because we must suppose that it is real things that provide those causal occasions. (ibid.)

The 'must suppose' is a crucial point. Realists may be tempted to read it as indicating that Kant's theory requires the existence of 'real things' in the sense of actual 'things-in-themselves'. I think, this would be the wrong reading. Rather, Kant speaks here of a need that arises in 'practical life', especially when we want to coordinate our actions with those of others. The thing-in-itself, Kant reiterates in many places (e.g., 1787; p.591, 610), is intended as a 'product of thought' (*Gedankending*) that serves as a 'heuristic fiction' (1787, p.799). To my mind, this covers *any* conception of an ontic reality that is structured in space and time. The fiction of such a reality, however, becomes necessary for the purpose of social interaction.

In a later essay on 'The ability to know' (*vom Erkenntnisvermögen*, the first section of his *Anthropology*, 1800) where he explains his approach to the senses, he returns to the notion of 'appearance':

Perceptions of the senses (empirical presentations with consciousness) can only be called internal *appearances*. It is not until the understanding that joins them and connects them by a rule of thought (which brings *order* into the manifold), that they become empirical knowledge, i.e., *experience*. (Kant,1800, *Werke*, Vol.vii, p.144)

Kant's term 'the manifold' (*das Mannigfaltige*) indicates another key concept. It is comprehensible only in conjunction with the basic presupposition of his theory, that space and time are the fundamental forms which human reason imposes on all experience. These forms are a priori because they are inherent in the functioning of reason. The 'manifold', then, is the raw material, the stuff on which constructive perception and reason can operate. William James called it, 'one big blooming buzzing confusion' (James, 1962, p.29). In present-day neurophysiology one would say, it is the totality of electrochemical impulses continuously generated by the sensory organs of the system. Even if one assumes that these impulses are caused by differences of an ontic substrate they cannot convey qualitative information, because qualitatively they are all the same.

Experience, thus, is what the thinking subject coordinates (constructs) out of elements of the manifold — and the fact that only certain things are constructed, and others not, is determined by the structure of reason, which Kant considers the primary topic of his transcendental philosophy. This philosophy is rightly called 'rational idealism'. It proposes a painstaking and ingenious model that reason constructs of itself and it reduces the view of the universe entirely to ideas. For whatever lies outside the domain of reason, Kant uses the term *noumenon*, and he affirms that, though the assumption of noumena is rationally necessary, they remain unknowable. Thus he returns to the position of the 'negative' theologians and pits his agnostic model against all the great philosophers who preceded him.

From the beginning of western philosophy, the knowledge human reason constructs was in some way thought to be related to an independent reality. For most of the thinkers the relation had to be some form of representation; an imperfect, nebulous representation perhaps, but still a picture that correctly rendered some aspects of reality. Berkeley and Vico realized that reason could not have the required access, so they circumvented the problem, each in his own way, by making the connection through God, the creator of both the real world and mankind. All idealist and rationalist philosophers, from Plato to Leibniz, drew in some way on the notion of God to prevent their systems from slipping into solipsism, i.e., the idea that the world has no existence outside the thinker's subjective mind.

Kant then systematically dismantled all previous attempts rationally to prove the existence of God. But he concluded that it is equally impossible rationally to prove the non-existence of God, and that the undeniable possibility of His existence should be sufficient for the believer (1787, pp.770, 781). As he did in the case of the thing-in-itself, he claimed that to assume an all-powerful creator of the world, is rationally necessary, but that this does not add to our knowledge, because it means no more than 'to assume a something of which we have no conception what it might be in-itself' (pp.725–6).

A Re-assessment of Causality

Throughout the nineteenth century science expanded in every conceivable direction and provided a vast seedbed for technology. The development of mechanical gadgets and machines flourished as never before. On the one hand, they profoundly modified the experiential world of human beings in the West and beyond; on the other, their practical success helped to reinforce the illusion that the theories from which they sprang were coming closer and closer to representing the world as it really is. Before the turn of the century, there were not only popular writers but also scientists who proclaimed that the important problems had been solved, and all that remained was the need to mop up a few details here and there (see Bernal, 1971; p.665). The once mysterious ways the world works had all been reduced to causal relations.

But there were other scientists — and some of the greatest among them — who did not share such facile optimism. Hermann von Helmholtz, for instance, who had been an attentive reader of both Hume and Kant, wrote:

> Not until later [in my life] did I make clear to myself that the principle of causality is, in fact, nothing else but the presupposition of the lawlikeness of all appearances of nature. (Helmholtz, 1881/1977)[16]

Causality, then, is part of the design that reason imposes on experience to make it understandable. But where does so particular a design come from? Hume suggested that it arose from the repeated contiguity of perceptions in the flow of experience. This idea was soon discredited by a simple observation: in our experience, day is contiguous with night, and night with day, yet it makes no sense to consider either the cause of the other. For Kant, the relation of cause and effect was a 'synthetic a priori' category, inherent from the outset in our thinking. He did not mean that it was innate or God-given in the sense of a Platonic idea, but rather that it was one of those heuristic fictions that reason needed in order to generate a rational picture of itself as the producer of understanding.

Such circularity is an inevitable characteristic not only of Kant's transcendental philosophy but of *any* attempt to construct a rational model of how we generate a coherent picture of the world from *within* our experience. It is the means that bridges gaps which the mystic fills with a poetic metaphor. The constructivist is well aware that circularity cannot be avoided — but he would like to reduce it to a minimum. In the case of causality, a plausible conceptual analysis was not supplied until much later, in Piaget's *Genetic Epistemology* (see Chapter 3).

New Fuel for Instrumentalism

For all instrumentalist approaches to knowledge, the most important event of the nineteenth century was the publication of Darwin's theory of evolution. William James was perhaps the first to make the relevant connection. In a brilliant essay, in which he pits Darwin's precise notion of selection against Spencer's hollow sociological assumptions, he says of the origin of new conceptions that they:

> . . . are originally produced in the shape of random images, fancies, accidental out-births of spontaneous variation in the functional activity of the excessively instable human brain, which the outer environment simply confirms or refutes, adopts or rejects, preserves or

destroys — selects, in short, just as it selects morphological and social variations due to molecular accidents of an analogous sort. (James, 1880, p.456)

On the following page he applies the idea to the scientific investigator, and it becomes clear that by the loose expression 'the outer environment' he was not referring to an independent, objective world:

To be fertile in hypotheses is the first requisite, and to be willing to throw them away the moment experience contradicts them, is the next. (ibid., p.457)

He is speaking of 'experience', not of a world as it might be in itself.[17] That this is so, can be gathered from many other statements he made in the context of his theory of pragmatism (e.g. James, 1907; p.49).

The notion that Hypotheses are maintained only if they find some confirmation in experience, was certainly not new. But that this was analogous to living organisms surviving in an environment, was revolutionary. Until then, hypotheses were thought to become theories and, through further confirmation, factual accounts or laws that were believed to represent an objective reality. Now the progress of science and human knowledge in general could be seen as a continuous evolution which, as an analogy, was in complete harmony with Darwin's biological theory.

This view was quickly spread and, as so often when philosophical ideas are simplified for distribution, it was encapsulated in a slogan. Pragmatism became known as the movement that proclaimed: 'Truth is what works'. Because one could say that what survives and is able to reproduce, works in biological evolution, it was assumed that also in the domain of concepts and ideas, the criterion was a simple and utilitarian one. James himself occasionally contributed to this interpretation, e.g., when he wrote of the 'pragmatistic view':

. . . all our theories are instrumental, are mental modes of adaptation to reality, rather than revelations or gnostic answers to some divinely instituted world-enigma. (James, 1907/1955, p.127)

However, 'confirmation in experience' is a far more complex matter when it involves conceptual structures rather than biological responses or attributes. The 'mode of adaptation' on the conceptual level is not the same as on the physical level of the organism (conceptual equilibration will be discussed in Chapter 3).

The German philosopher and sociologist Georg Simmel remarked early on that the evolutionary approach

eliminates the dualism of an independent truth in-itself and . . . experience or selection concerning the practical interaction with the world as it comes to be known — because the experience of the effects of one's actions at the same time creates truth. (Simmel, 1895, p.44)

In other words, the requirement that knowledge be called true only if it reflects a real world, is relinquished for the requirement that it be found conducive to the attainment of our goals in the world as we experience it. One problem with this view arises once it has become clear that the way we experience the world is dependent on the hypotheses and the knowledge that help us to conceptualize our experiential environment. This is what Heisenberg meant when he said that the further natural scientists look into nature, the more they realize that what they are seeing is a reflection of their own concepts (see Chapter 1).

In spite of this problem, the movement of evolutionary epistemology that developed around the work of Konrad Lorenz, has gained considerable momentum, especially in the extended form given it by Donald Campbell, who characterizes it as 'hypothetical critical realism'. He agrees with Lorenz, that the concepts of space, time, and causality are not, as Kant thought, a priori elements of human reason, but rather the result of living organisms' adaptation to the universe. But he claims that modern physics 'provides a much finer grained view of reality'.

The *Ding an sich* is always known indirectly, always in the language of the knower's posits, be these mutations governing bodily form, or visual percepts, or scientific theories. In this sense it is unknowable. But there is an objectivity in the reflection, however indirect, an objectivity in the selection from innumerable less adequate posits. (Campbell, 1974, p.447)

The flaw in assuming such a 'reflection' of objective reality is that there is no reason to believe that any evolved structure, be it physical, behavioural, or conceptual, that proves viable (i.e., adapted) at the moment, is necessarily on the way to the best possible adaptation. The natural selection that preserved what lives today could choose only among the variations that accidental changes had actually produced. On the conceptual level, then, the 'innumerable less adequate posits' that Campbell mentions, leave out the far more innumerable ones that were never tried because they were incompatible with some basic principle that seemed indispensable at the time. Besides, the notion that the adaptedness of organisms provides a glimpse of the structure of Nature as it is, hardly jibes with the biologist's finding that the vast majority of species that were evolved and survived for millions of years, were nevertheless extinguished at some point.

However, there is a more basic logical flaw in the premises of evolutionary epistemology. Lorenz wrote: 'Adaptation to a given condition of the environment is equivalent to the acquisition of information *about* that given condition' (1979, p.167). This is the primary assumption of his school of thought, and it is unwarranted. The biological notion of fitness or viability does not require that organisms or species have information about or share properties with an independently 'given' environment. Adaptation merely requires that they avoid points of friction or collision. Whatever has passed through the sieve of natural selection might know that it has passed, but this does not provide any indication of the structure of the sieve. Both in the theory of evolution and in constructivism, 'to fit' means no more than to have passed through whatever constraints there may have been.

Hypotheses and Fictions

From 1876 to the first decade of our century, Hans Vaihinger, the founder of the German Kant Society, worked on a monumental work he called 'The philosophy of as if' (*Die Philosophie des Als Ob*, 1913)[18]. In it, starting from Kant's critical work, he develops his theory of 'fictions' that pertains to the entire field of human intellectual production and proposes a new way of looking at philosophy. He mentions Bentham as a forerunner and then proceeds to develop from the 'as if' principle a vast analytical enterprise that covers all areas but focuses in particular on philosophy and science. Rather than base his investigation, as Bentham did, on the common use of language, he follows Kant and analyses the possibilities of conceptualization. This leads him to accentuate the extremely important distinction between 'heuristic fictions' (a term that stems from Kant) and 'hypotheses'.

The way he distinguishes the two concepts sounds quite simple but is bound to be misunderstood if the reader does not take into account its derivation from Kant's theory of rational knowledge.

> An hypothesis, as we have seen, must be verifiable. It will be definitely included in the stock of scientific ideas when it has been found to be true, i.e., verified . . . A fiction cannot be confirmed by experience, but it can be justified by the service it renders to science . . . Once justified, the fiction will be admitted as a useful member to the domain of ideas. When it helps a thought computation to yield a *practically* useful result, as for instance, when the method of infinitesimals makes a curve computable, when an artificial or fictitious partitioning yields a *practical* order, then such auxiliary ideas are justified. . . . Just as the hypothesis is submitted to a test of the experiential reality of what was hypothesised, so the fiction is tested as to the practical usefulness and appropriateness of what it invented. (Vaihinger, 1913, pp.610–11)

The 'verification' referred to at the beginning of the quoted passage is not intended ontologically, but, as the author makes clear in what follows, he intends confirmation by experience. Though Vaihinger created a considerable stir among European thinkers, he was all but ignored by English-speaking philosophers. His notion of useful fictions, however, has recently cropped up under another name. Gregory Bateson, in his well-known and often quoted 'Metalogue: What is an instinct?' (1972a), speaks of an 'explanatory principle' which, like gravity, can explain 'anything you want it to explain'. Bateson's way of distinguishing explanatory principles from hypotheses is not as explicit as Vaihinger's, but it links the idea of useful fiction to that of the cybernetician who constructs a conceptual or mechanical model to substitute for something that is inaccessible. He explains it to his daughter as follows:

> F: . . . You see, an hypothesis tries to explain some particular something but an explanatory principle — like 'gravity' or 'instinct' — really explains nothing. It's a sort of conventional agreement between scientists to stop trying to explain things at a certain point.
> D: Then is that what Newton meant? If 'gravity' explains nothing but is only a sort of full stop at the end of a line of explanation, then inventing gravity was not the same as inventing an hypothesis, and he could say he not *fingo* any hypotheses.
> F: That's right. There's no explanation of an explanatory principle. It's like a black box. (Bateson, 1972a, p.39)

As Vaihinger illustrated in many ways, we could not even begin to construct a picture of the world, without employing fictitious entities and relations. However, what he showed to be crucial for the proper understanding of human understanding, is that these fictions be recognized as tools for the rational organization of experience and not mistaken for phenomena that are real in the sense that they themselves could be experienced.

The Foundation of Language Analysis

During the forty years straddling the turn of the century, there emerged quite a few people who broke away from supposedly established ideas and opened new perspectives. The upheaval is no doubt most obvious in the visual arts, but it became apparent also in literature, music, and philosophy. I am not competent to judge to what extent the surge was due to the revolution in physics, but from a constructivist point of view, authors such as Poincaré, Duhem, Mach, and the mathematician Brouwer, contributed to preparing the way. Some of their ideas will crop up in later chapters.

One thinker, however, is of particular importance for the constructivist approach because he gave the study of language a new foundation from which it could embark on a scientific track that was different from traditional philology. Ferdinand de Saussure has become known as the father of modern linguistics, although he himself never published a book. This is not the only oddity about his work. If one comes to Saussure because of the frequent references to him in the works of other authors, one cannot help being struck by the fact that few modern linguists have actually taken up what is, in a philosophical sense, the most important principle he laid down, namely that the meaning of words is to be found in the minds of speakers, rather than in the domain of so-called real objects.

We owe a compendium of Saussure's theory to two of his students who compiled an extremely interesting and readable book from their own and other students' notes, and, most importantly, their teacher's lecture notes (de Saussure, 1916/1959).

The feature of Saussure's investigation that distinguishes it from philology and much of later linguistics, is that he does not begin by analysing a vocabulary or grammatical rules, but rather by examining how language functions. When two people speak to each other, he notes, both utter sounds and both hear the sounds the other utters. He shows this in a diagram with two speakers linked by two arrows forming a circuit.

Suppose that two people, A and B, are conversing with each other. Suppose that the opening of the circuit is in A's brain, where mental facts (concepts) are associated with representations of the linguistic sounds (sound-images) that are used for expression. A given concept unlocks a corresponding sound-image in the brain; this purely *psychological* phenomenon is followed in turn by a *physiological* process: the brain transmits an impulse corresponding to the image to the organs used in producing sounds. Then the sound waves travel from the mouth of A to the ear of B: a purely *physical* process. Next, the circuit continues in B, but the order is reversed: from the ear to the brain, the physiological transmission of the sound-image; in the brain, the psychological association of the image with the concept. If B then speaks, the new act will follow — from his brain to A's — exactly the same course as the first act and pass through the same successive phases, . . . (de Saussure, 1959, p.11–12)

This explanation is both simple and fundamental. It provides a model of the mechanics of communication that illustrates two things.

1   The two-way correspondence between sound-images and concepts, which is in fact the *semantic* connection between a word and its meaning, is the result of psychological association. Psychological associations, however, can be formed only by an individual in his or her subjective experience (see Chapter 7).

2   Since no individual's experience can cover all the situations that have given rise to the associations (i.e., the semantic connections) that have been formed by the social group as a whole, the collective sense of the word 'language' (Saussure's *langue*) requires an abstraction that even

> a diligent observer of a great many linguistic interactions can only
> hope to approximate.

If one accepts this analysis, the notion collapses that every child growing up in a linguistic community will automatically associate the sound-images it perceives with concepts that are *shared* with the entire community. Instead, learning the language will be seen as a never ending process of adaptation of one's own concepts, governed by the need and the wish to establish mutually compatible associations to the speech sounds one is hearing and producing.

The expression 'shared meaning' is therefore a little misleading. Awareness of the ambiguity of the word 'to share' may help to clarify this. It is one thing, to share a car, but quite a different thing, to share a bottle of wine. In the first case, two or more individuals are using one and the same car; in the second, none of the wine drunk by one person can be drunk by another. Sharing a meaning is a little like the second example, but not at all like the first. We cannot share our experience with others, we can only tell them about it, but in doing so, we use the words that *we* have associated with it. What others *understand* when we speak or write is necessarily in terms of the meanings their experience has led them to associate with the sound images of the particular words — and their experience is never identical with ours.

Given that language users as a rule achieve a great deal of linguistic compatibility with the others of their group, they easily come to believe that the words they use actually refer to objects in a real world and that, therefore, language does provide a description of things beyond individual experience. The implicit reasoning that leads to this illusion is something like: if so many refer to the same things, the things must be real. But this overlooks the way in which each language user constructs meanings, and that these meanings had to be adapted to the others' use of words and thus modified the practice of segmenting and talking about experience.

In discussing the relation between language and reality, Richard Rorty speaks of the temptation to 'think that there is some relation called "fitting the world" or "expressing the real nature of the self" ' (Rorty, 1989, p.6). I hope to have shown that succumbing to that temptation would lead to an altogether untenable position, even if one interprets the 'fitting' in the constructivist way as 'being compatible', rather than 'matching', which would be the realist's interpretation. The only fit we can assess is a fit with the world as we experience it.

Conclusion

It has no doubt become clear that this trot through the history of ideas is the subjective presentation of pieces an eclectic reader has collected in his attempt to construct a relatively coherent, non-contradictory model of knowing. It does not claim to present the truth about what the authors of the quoted passages intended, but only a viable reading. I do not believe that any amount of hermeneutical research can produce a *true* replication of the concepts historical thinkers had in mind. I have therefore chosen to interpret their texts as best I can from my point of view. I hold with the French poet (and mathematician) Paul Valéry, who said:

> I have already explained what I think of *literal* interpretation; but one cannot insist enough on this: *there is no true meaning of a text.* No author's authority. Whatever he may have wanted to say, he wrote what he wrote. Once published, a text is like an implement that everyone can use as he chooses and according to his means: it is not certain that the maker could use it better than someone else. (Valéry, 1957, p.1507)

I set out to substantiate the claim that reason could not deal with the mystical and its wisdom. The pre-Socratics already argued that a reality independent of the human ways of knowing was not accessible to us, because we cannot step out of our ways of knowing. This was a purely logical limitation. Early Christian theologians added another argument: because our concepts are formed by abstraction from experience, they cannot capture anything that lies beyond our experience. The medieval mystic, John Scottus Eriugena, then anticipated both Vico and Kant in saying that reason can know and understand only what it itself has made according to its own rules.

The birth of modern science in the Renaissance brought forth the suggestion that scientific knowledge was instrumental and should therefore be kept apart from the mystical, which was timeless. But the rediscovery of Pyrrhon's early school of scepticism encouraged some to use the sceptics' arguments against certain knowledge to question the dogma of the Church. Descartes intended to stop this by demonstrating that there were, indeed, things that could be known with certainty. His method of radical doubt, in the end, only confirmed the sceptics' position.

Each of the three famous empiricists provided basic insights into the process of knowledge construction. Locke spoke of reflection upon mental operations as a source of ideas; Berkeley noted that time, succession, number, and other indispensable concepts, are mental constructs; Hume, explained the active generation of relational concepts by acts of association.

Berkeley's contemporary, Vico, produced the first explicit formulation of a constructivist approach — human reason can know only what humans themselves have made; and, more importantly perhaps, he suggested a way to distinguish the language of the mystic by its irreducible metaphors from the language of reason which is anchored in experience.

Kant's analysis of the rational domain then confirmed the inaccessibility of anything posited beyond the reach of experience and maintained that the world we understand and live in becomes real to us, because we complete the picture by means of rational heuristic fictions.

Following this lead in Kant's work, Vaihinger produced his 'Philosophy of as if', an analysis of western intellectual culture and science in terms of useful fictions. Although he occasionally refers to the problem of children's construction of concepts, his approach is on the whole a philosopher's who considers knowledge to be more or less static. The fact that he says little about *how* fictions are built up, is the main reason why he cannot properly be called a constructivist.

The concepts of variation and (natural) selection, taken from Darwin's theory of evolution, opened the possibility of substituting the notion of adaptedness for the philosophers' traditional notion of truth as a correct, or a least approximately correct, representation of objective reality. However, the question of what the constraints relative to which adaptation occurs are, is answered differently by contemporary schools. Evolutionary epistemologists, for example, tend to reduce all knowledge — including the elementary concepts of space and time — to biological adaptation in the service of survival. For them, the theory of evolution is an unquestionable ontological given and thus the basis for the illusion that the products of adaptation provide positive information about the constraining world. I call it an illusion because none of the 'critical realisms' that have been based on the evolutionary principle is able to show *how* a cognizing subject might turn its notion of adaptedness into *knowledge* of reality.

Finally, de Saussure characterized language as

> a system of signs in which the only essential thing is the union of meanings and sound-images, and in which both parts of the sign are psychological. (de Saussure, 1959, p.15)

Since this union has to be created by every language user on the basis of his or her individual experience, the meanings we attribute to words cannot be anything but subjective. This eliminates the traditional philosophical 'Theory of Reference', which is based on the notion that words refer to things-in-themselves. Instead, words can now be thought to refer to whatever abstraction from experience, i.e., whatever meaning, the individual language user happens to have made.

The notion of communication arises from the assumption that organisms who live in groups and have the capability of abstracting images and ideas from their experiences, will make many such abstractions in situations where they are in the company of others — which leads them to the assumption that the others have made the same abstractions as they themselves. Once they associate sound-images of words with their ideas, they will come to believe that the meanings of words are the same whenever their interactions with others show them to be compatible. Since such compatibility is crucial in many forms of necessary collaboration, the members of a community will do their best to make their meanings compatible with those of others.

Our meanings, then, may be modified and adapted to common usage in the constant non-linguistic and linguistic interactions we have with others;

but the result of such adaptation will at best achieve a relative compatibility, never identity.

From all this, with the help of Piaget's theory of cognitive development (see Chapter 3), radical constructivism formulated its fundamental principles:

1 • Knowledge is not passively received either through the senses or by way of communication;
 • knowledge is actively built up by the cognizing subject.
2 • The function of cognition is adaptive, in the biological sense of the term, tending towards fit or viability;
 • cognition serves the subject's organization of the experiential world, not the discovery of an objective ontological reality.

The last of the four points seems the most difficult to accept. Even Kant, who had so many brilliant ideas about the ways and means of conceptual construction, was not inclined to give up the search for ontological truth. Some serious critics of radical constructivism are driven by the same attachment.[19] They refuse to consider that this theory of knowing is intended as a tool that should be tested for its usefulness rather than taken as a metaphysical proposal.

Notes

1 Since I have no Greek, the translations I give of the pre-Socratics are based on the German translations of Hermann Diels (1957) and the English ones by W.K.C. Guthrie (1962, 1971).
2 The idea was also picked up by the eighteenth-century philosopher Berkeley (1732/1950, p.166).
3 Translation by Sheldon-Williams, quoted in R. Kearney (1985).
4 Note that throughout this text the word 'model' is used, as in cybernetics, to refer to a physical or conceptual structure invented to simulate the behaviour of a 'black box', i.e., something that behaves in unexpected or interesting ways, but whose insides are not accessible to observation.
5 The argument whether Galilei continued to believe, like Plato, in the transcendence of mathematical laws or, late in his life, came closer to the view suggested by Bellarmino, is still going on (see Belloni, 1975).
6 Since the edition of Berkeley's collected works by Luce and Jessop (1950), the notebook is called 'Philosophical Commentaries'; earlier it was known as the 'Commonplace Book'.
7 I mentioned in Chapter 1 that Bishop Berkeley also developed a metaphysics in which the real world was held together by God. Though Locke and Hume had no official link to the Church, they had no sympathy for atheism and devoted much time and space to reconciling their sceptical view of rational knowledge with the accepted faith in the revelation of God's reality. Constructivism, as I stated at the beginning, suggests an approach to rational thought and therefore neither engages in nor denies metaphysical considerations.

8　Bentham used the term 'entity' for what today we might call an 'item', i.e., anything whatever.

9　Note that in present-day English, 'it's true that . . .' and 'it's a fact that . . .' are also used interchangeably.

10　I refer to the 1850 edition of Vico's treatise, which contains the Latin text as well as Pomodoro's translation, giving the original numbers of chapters, paragraphs, and propositions.

11　Those who have read Piaget will be startled by this remarkable anticipation of Piagetian ideas.

12　'Every metaphor', he says, 'is a fable in brief' (1744, par.404).

13　In spite of the clarity of this distinction, Vico himself did not waver in his religious faith and devoted a great deal of time and words to metaphysics. In a manner parallel, but not identical with that of Berkeley, he attempted to give eternal validity to man's rational constructs.

14　Kant uses the word *Erkenntnis*, which contains the German root of cognizing rather than that of knowing; hence I translate it as 'cognition'.

15　The word *Vorstellung* is a key term in Kant's philosophy. When it is translated as 'representation', this is bound to mislead, because this English word suggests that there is an original that is being represented. The end of the quoted sentence makes clear that Kant uses the word as it is normally used in German, namely indicating something one presents to oneself spontaneously and not as the copy of something else.

16　This was written in 1881 as an addition to Helmholtz' 1847 treatise on 'The conservation of force'. It can be found on p.180 of his Epistemological writings, 1977. Historically, it is of interest that the passage was quoted in one of the last lectures of the course on 'The physical foundations of the natural sciences' by Franz Exner (1919), whom Erwin Schrödinger later cited as one of his most influential teachers.

17　On the relation between experience and reality, James has made a statement that is as profound as it is subtle: 'Everything real must be experienceable somewhere, and every *kind* of thing experienced must somewhere be real' (1912, p.159; my emphasis). I have italicized 'kind' because it would be easy to overlook the distinction James is making: Like Berkeley, he calls only those things 'real' that can be experienced somewhere; and kinds of things, i.e., the concepts we have abstracted, must be based on things that are 'real' in the sense he has defined — otherwise they are hollow or, as Vico would say, 'poetic metaphors'.

18　An excellent, slightly abridged translation by C.K. Ogden, reprinted by Barnes and Noble (New York, 1968) makes the sixth edition of this work accessible to English readers.

19　Recently it has been suggested that radical constructivism is contradictory because it attacks realism and at the same time assumes a realist position by admitting that an ontological reality may constrain human action (e.g., Matthews, 1992, p.186). In the usual language of philosophers, 'realists' are those who believe that they can obtain knowledge of a world as it is in itself. This I deny, and admitting 'ontic' constraints does not contradict it, because while they may determine what is impossible, they do not determine the ways of acting and thinking that can be constructed within them.

Chapter 3

Piaget's Constructivist Theory of Knowing

It is a difficult task to glean a coherent theory of cognitive development from Piaget's enormous body of work. Over a period of seventy years, Piaget published eighty-eight books, hundreds of articles, and edited countless reports of research that had been carried out under his supervision.[1] His thinking and his ideas never ceased to develop, to branch out, and to spiral into new formulations which, in his mind, continuously expanded and modified what he had expressed in earlier writings. As a result, it requires considerable effort to sort out what seems to have remained the same and what was modified in the course of those decades. Those who venture to summarize Piaget's ideas on the basis of two or three of his books have a limited perspective. They inevitably remain unaware of implications that cannot be grasped except from other parts of his work. Unfortunately, there are countless psychology textbooks and critical journal articles that fail in this respect. At best they provide an incomplete view of Piaget's theory, at worst they perpetuate distortions of his key concepts. Many summarizers and critics, moreover, seem to have missed, or simply disregarded, the revolutionary approach to epistemology that Piaget developed as the basis of his investigations. This second failing is the more serious. Without the understanding that Piaget quite deliberately stepped out of the western philosophical tradition, it is impossible to come to a comprehensive view of his theory of knowing and the model he built to explain how children acquire knowledge.

Piaget is not easy reading. Although he never ceased to praise the virtue of 'decentration' — the ability to shift one's perspective —, he himself, as a writer, did not always try to put himself into his reader's shoes. I feel that writing often was for him, as for many original thinkers, part of working out his ideas for himself. His untiring efforts to express his thoughts in the greatest possible detail do not always help the reader's understanding. Yet, I never doubted that it was worth trying to overcome those difficulties, for the effort has led me to a view of human knowing that no other source could have provided.

For six or seven years I concentrated almost exclusively on Piaget; and I have sporadically returned to his writings for almost two decades since then. Yet I want to emphasize that what I lay out here is the sense that *one* rather diligent reader has extracted. It is certainly not the only possible interpretation, let alone an official one. But it is an interpretation that I have found cogent and extremely useful in a variety of applications. This , however, does not make it any less subjective.

There are at least half a dozen concepts that have to be characterized with a certain precision if we want to arrive at a coherent interpretation of Piaget's theory. The task of characterizing someone else's concepts is necessarily a conjectural one. One cannot enter another's head to examine what conceptual structures he or she has associated with certain words. As readers of Piaget's writings, therefore, we can only conjecture what a given word meant to him when he used it. As we come across the word again and again in his works, we can try to modify or reconstruct our supposition in the hope of arriving at an interpretation that fits, if not all, at least a large number of occurrences. In principle, this is the process of hermeneutics, the art of unravelling the original meaning of texts. It should be clear that there can be no absolute answers. The reader's attempt to construct for each word a constant meaning that might fit all the encountered contexts can yield only relative results. On the one hand, the notion of fit is inevitably a relative one and, on the other, it is based on the assumption that meanings are constant for a given author. This assumption is obviously an unlikely one in the case of an author who, like Piaget, has used some of his key words for many decades during which his thinking continued to expand. Yet, I am convinced that the direction of his search remained unchanged throughout. The interpretations and definitions I am giving here are the ones that make sense to me in the light of those of Piaget's works, and certain passages in them, that I consider central.

The Biological Premise

Piaget was unquestionably the pioneer of the constructivist approach to cognition in this century.[2] This approach was unconventional when he developed it in the 1930s, and it still goes against the generally accepted view today. It is also an uncomfortable approach, because it requires drastic changes of certain fundamental concepts that have been taken for granted for thousands of years. Among these fundamental concepts are 'Reality', 'Truth', and the very notion of 'what knowledge is' and 'how we come to have it'.

To explain why and how Piaget came to such a drastic break with the western philosophical tradition, we have to look first of all at the starting point of his intellectual career. For the rest, I want to stress that my underlying assumption in this interpretation is that it was Piaget's aim to produce as coherent a model as possible, of human cognition and its development. Although from the outset he had a clear idea of the direction he was going to take, he could not possibly have foreseen all the steps. His model did not grow in a straight line but rather like a tree, some of whose branches wither, as the central ideas are developed further. This means that I shall disregard statements in his earlier writings that appear to be in flat contradiction with later ones.[3]

In one of the few autobiographical passages he wrote, Piaget recounts that he decided 'to consecrate my life to the biological explanation of knowledge' (Piaget, 1952b, p.240). It would be difficult to overrate the importance of this statement. To consider cognition as a biological function, rather than the result of impersonal, universal, and ahistorical *reason*, constitutes a radical break with the philosophers' traditional approach to epistemology. To begin with, it shifts the focus from the ontological world as it might be, to the world that the organism experiences.

As far as I know, Piaget had no contact with the work of Jakob von Uexküll, but there is a certain similarity in some of the two thinkers' ideas. What the German biologist called *Merkwelt*, the world of sensing, and *Wirkwelt*, the world of acting (von Uexküll and Kriszat, 1993), is included in Piaget's notion of the 'sensorimotor level'. Both authors had been profoundly influenced by Kant's insight that whatever we call knowledge is necessarily determined to a large extent, if not altogether, by the knower's ways of perceiving and conceiving.

Piaget himself described the goal of his undertaking in his introduction to *The Essential Piaget* (Gruber and Vonèche, 1977):

> The search for the mechanisms of biological adaptation and the analysis of that higher form of adaptation which is scientific thought, the epistemological interpretation of which has always been my central aim. (Piaget, in Gruber and Vonèche, 1977, p.xii)

That the acquisition of knowledge was 'adaptive', had been suggested by James, Simmel, and others around the turn of the century, but Piaget saw early on that adaptation in the cognitive/conceptual domain was not the same as the physiological adaptation of biological organisms. On the level of cognition, he realized, it was not a straightforward matter of survival or extinction, but rather of conceptual equilibration. It is important, therefore, to keep in mind that when he speaks of 'that higher form of adaptation', the mechanisms he is looking for are mental and not biological as in the ordinary use of that term.

It was this search for the mechanisms of cognition that motivated Piaget's interest in children. By observing the interactions of infants and growing children with their environment, he intended to isolate manifestations of cognitive processes in order to arrive at a generalizable model of cognition and its ontogenesis. From the traditional philosophers' point of view, whatever such an enterprise could yield would be a 'genetic fallacy', because for them, *knowledge* has to be timeless and immutable and can never be justified by the history of its generation. Consequently, most philosophers felt justified in disregarding whatever Piaget said and wrote. Psychologists and the general public, on the other hand, concluded that he was a child psychologist, because his texts frequently referred to developmental phenomena in children. Given this perspective, they did what they could to adjust his ideas so that they could be fitted into the psychological tradition. This often quite unconscious effort was probably the main reason for the colossal misinterpretations that are rampant in the literature.

Active Construction

A typical example is the following. In many places (e.g., 1937, p.10; 1967a, p.10; 1970a, p.15), Piaget explains that, in his view, knowledge arises from the active subject's activity, either physical or mental, and that it is goal-directed activity that gives knowledge its organization.

> . . . all knowledge is tied to action, and knowing an object or an event is to use it by assimilating it to an action scheme . . . (Piaget, 1967a, pp.14–15)

> . . . to know an object implies its incorporation in action schemes, and this is true on the most elementary sensorimotor level and all the way up to the highest logical-mathematical operations. (ibid., p.17)

The concept of 'action scheme' is central in Piaget's theory of knowledge, and I shall explain it in detail later on. That it was widely misunderstood, is due above all to the fact that he derived it explicitly from the biological notion of 'reflex'. Action schemes were therefore tacitly interpreted by many readers as stimulus–response mechanisms. This made traditional psychologists feel comfortable because it allowed them to classify Piaget's theory as an 'interactionist' one — a somewhat complicated interactionism, to be sure, but certainly not a revolutionary doctrine that would shake their fundamental belief in universal objects in a real environment with which living organisms are supposed to interact. This misinterpretation simply confirmed the notion that interaction provides the intelligent organism with knowledge, and that this knowledge, through further interaction, becomes better, in the sense that it comes to reflect the environment more accurately. Thus, although Piaget frequently spoke of 'construction', he could be accepted as a somewhat idiosyncratic developmental theorist, and the psychologists' peace of mind was saved.

Once that view was established, only a direct contradiction might have disrupted it. But explicit contradictions of our age-old common-sense notion of knowledge and the world are difficult to find in Piaget's works. Whenever he says, for instance, that knowledge must not be thought of as a picture or copy of reality (and he says that often enough), it is easy to mistake this for a conventional admonition that a cognitive organism's picture of the world would necessarily be incomplete or somewhat distorted. Any realist will read it as such, rather than take it as an integral part of Piaget's view that knowledge, of its nature, cannot have any iconic correspondence with an ontological reality.

Piaget's position can be summarily characterized by the statement: 'The mind organises the world by organising itself' (1937, p.311). The cognitive organism shapes and coordinates its experience and, in doing so, transforms it into a structured world.

> What then remains is construction as such, and one sees no ground why it should be unreasonable to think that it is the ultimate nature of reality to be in continual construction instead of consisting of an accumulation of ready-made structures. (Piaget, 1970b, pp.57–8)

Almost none of Piaget's writings could be fully understood without taking into account this revolutionary perspective. Yet it is difficult to remain aware of it, because Piaget only rarely refers to the relation between knowledge and reality or reminds the reader that in his model the 'real' is always the experiential world.

The focus of his work was and remained throughout his long life the design of a viable model of how we manage to construct a relatively stable, orderly picture from the flow of our experience. That he achieved this to a far greater extent than anyone else, is due to several things: his refusal to accept dogmatic explanations, his unfaltering energy in asking new questions, his good fortune in finding an independently brilliant yet cooperative and empirically oriented colleague in Bärbel Inhelder, and the passionate explorer's attitude that he characterized in retrospect when he said:

> At the end of a career it is better to be ready to change, rather than condemned to repeat oneself. (Piaget, 1976b)[4]

Beginnings

Piaget embarked on a research career a good deal earlier than most scientists. Around 1907, when he was barely 11-years old, he observed an albino sparrow in a public park near his home in Neuchâtel. He wrote a brief note about it, and sent it to a natural history journal. The note was published, and on the strength of it he was allowed to spend his free time after school in helping Monsieur Godet, the director of the local natural-history museum, with the sorting of some collections. Growing up on the shores of lake Neuchâtel, he had already become interested in fresh-water molluscs and Paul Godet happened to be an expert in this field. It was a marvellous apprenticeship for the young Piaget.

In 1911, when Paul Godet died, the schoolboy (as Piaget wrote in an autobiographical sketch) knew enough about molluscs:

to begin publishing without help (specialists in this branch are rare) a series of articles on the molluscs of Switzerland, of Savoy, of Brittany and even of Columbia. This afforded me some amusing experiences. Certain foreign 'colleagues' wanted to meet me, but since I was only a schoolboy, I didn't dare to show myself and had to decline these flattering invitations. The director of the *Musée d'histoire naturelle* of Geneva, who was publishing several of my articles in the *Revue Suisse de Zoologie*, offered me a position as curator of his mollusc collection . . . I had to reply that I had two more years to study for my baccalaureate degree, not yet being a college student.[5] (Piaget, 1952b, pp.238–9)

In retrospect, one might say that the study of molluscs determined Piaget's intellectual career. In his minute observations of these creatures he noticed that their shells differed in shape according to their location in still or in running water. It was a clear case of adaptation. But by transplanting some of the molluscs from one environment to the other, he discovered that the shape of their shells seemed not to be due to phylogenetic but rather to ontogenetic adaptation. This difference intrigued him — so much apparently, that he spent the rest of his life studying the living organism's capability of ontogenetic adaptation in its most impressive manifestation on the human level, namely the capability of *knowing*.

" A theory of the knowing mind

The Construction of Experiential Reality

Since Piaget's theory is, in fact, a theory of the knowing mind, the key terms constitute a closely knit conceptual network and, in one way or another, are all linked and interdependent. To isolate each one in order to explain and define it in separation from the others, is not an ideal way of proceeding — but I know of no other possibility. Language is a linear affair and linguistic explanation requires that things be arranged sequentially, one after the other, no matter how intricate their mutual dependencies might be in a given author's mind or in the fabric of our own living experience. This basic problem of presentation is nowhere better illustrated than in Piaget's fundamental work *La construction du réel chez l'enfant* (1937); which also is a good place to begin an exposition of his theory.

This early book is an attempt to show that human infants can construct for themselves the reality they experience. Indeed, they must do this, *whether or not* we assume that such a reality exists independently. The book, of course, does not detail the construction of any particular infant's reality with mummy, daddy, teddy bear, and pot, but it shows *how* the basic concepts that constitute the essential structure of any individual's reality can be built up without the assumption that such a structure exists in its own right. This is a cornerstone of Piaget's theory and the most important difference between it and all traditional theories of knowledge. As a direct consequence of his maxim that 'knowledge is a higher form of adaptation', he relinquished the notion of cognition as the producer of representations of an ontological reality, and replaced it with cognition as an instrument of adaptation the purpose of which is the construction of viable conceptual structures.

The constructive activity during the first two years of life lays the foundation of what will become the child's experiential world: it forms the essential scaffolding for all further constructing. As the child's living experience expands, layer upon layer of conceptual constructs is built upon the foundation. At any subsequent stage of development, therefore, it is difficult, if not impossible, introspectively to retrace the path of the early construction or to change the concepts that were its immediate results.

The first eighty-five pages of *The Construction of Reality in the Child* describe the development of the notion of 'object'. There are two consecutive phases in this development. The first leads to the establishment of object concepts in the sense that the infant coordinates (associates) sensory signals of the 'perceptual' kind that happen to be recurrently available at the same time in its sensory field (the 'locus' of raw material that Kant called 'the manifold').[6] These concepts could be described as routines for the reconstruction of a particular object of interest, whenever its sensory components are available. Their successful composition may then serve as trigger for a specific activity that has been associated with the object. In that case, an observer might say that the child recognizes the object, although the child may still be unable to conjure up a *re-presentation*, that is to say, a visualized image of the object, when the relevant sensory material is *not* actually available (see Chapter 5).

The second developmental phase can occur only when the infant has reached the stage of 'deferred imitation' (Piaget has called this the sixth stage of sensorimotor development, which usually falls within the time frame between the eighteenth and the twenty-fourth month of age). Deferred imitation refers to the child's ability to run through a sequence of physical actions when the perceptual situation that originally led to the coordination of the sequence is not actually present. When the deferred execution does not involve a motor activity but the conceptual coordination of a previously constructed object, it produces a *re*-presentation.

Unfortunately, Piaget only occasionally spells the word 're-presentation' with a hyphen (e.g. in his *La formation du symbole chez l'enfant*, 1945). To my mind, the hyphen is essential because Piaget uses the word in a sense that is very different from that intended by contemporary philosophers. For Piaget, *re*-presentation is always the replay, or *re*-construction from memory, of a past experience and not a picture of something else, let alone a picture of the real world.

I see a useful analogy to these two phases in the acquisition of vocabulary involved in the learning of a language. No matter what your level of proficiency in a natural language, there will be words that you know when you

replay or re-construction from memory

59

read or hear them, yet they are not available to you when you speak or write. You are able to recognize them, but not to *re*-present them to yourself spontaneously. To some extent this is the case in anyone's first language, but it is usually more noticeable in a second language, where one's reading knowledge — which does not involve the problems of pronunciation — is a good deal greater than one's speaking knowledge.

The ability to re-present objects to oneself is linked to language acquisition also in a very direct way. As long as words are used with immediate reference to the situation in which they are uttered, the speaker will be satisfied that the receiver has 'understood' the utterance, if the receiver's reaction is compatible with the speaker's expectation. This sort of 'understanding' is manifested, for instance, by a dog which sits down whenever the master says 'sit!'. The dog does not need to have a re-presentation of the meaning of the word 'sit'. Obedience to the command merely requires that the dog has associated the experience of hearing the sound patttern of this particular word with the act of sitting down. In contrast to this, if I said 'last night I sat down on a wet bench in the park', you or any other proficient speaker of the English language would not react with a specific motor pattern, but you would produce a mental re-presentation of some experience that, from your point of view, fitted the meaning of that sentence. That is, my utterance would be understood as a sequence of symbols that have to be interpreted conceptually, rather than as a signal associated with a physical reaction. (see Glasersfeld, 1987).

The ability to produce re-presentations of objects is one of the two essential ingredients of the development of 'object permanence', and this is probably the ontogenetically first context in which re-presentations appear. Later they become the indispensable basis for the most important conceptual activities, such as the presentation of hypothetical situations, hypothetical goals, hypothetical perturbations, and thus for the making of reflective abstractions from experiences that have not actually taken place on the sensorimotor level. I shall return to this in the chapter on reflection and abstraction.

Individual Identity

The second essential ingredient in the construction of permanent objects is the notion of individual identity. Prior to that notion, a comparison between a present experience and the re-presentation of an object will yield a judgment of either difference or sameness. The notion of individual identity complicates that issue because it introduces the possibility to construct two kinds of sameness. On the one hand, there is the sameness of two experiential objects that are considered the same in all respects that have been examined (as in assimilation); this can be called 'equivalence'. On the other hand, there is now the sameness of two experiences that are taken to be two experiences of one individual object.

The difference between the construction of equivalence as the basis of classification, on the one hand, and the construction of permanence as the basis of individual identity, on the other, springs from assigning perdurance to different items.[7] In the first case, the set of characteristics which, as a group, differentiate a particular experiential item from all other constructs, is abstracted and maintained (given perdurance) for future use. It constitutes the template or prototype to which experiences can be assimilated as members of the class. This procedure is the basis of all classification and categorization.

The concept of 'object permanence', on the other hand, is an abstraction from the second type of sameness. It characterizes the situation where a child considers the object it is perceptually constructing at the moment, to be the *identical* (self-same) individual it experienced at some prior time.[8] Perdurance is now attributed to the object whether or not it is actually being experienced. Consequently, the item can be said to 'exist'.

The element of individual identity is indispensable for the construction of several other fundamental concepts, such as state and change, process and motion, space, causality, and time (see Chapter 4). To each of the last three of these, Piaget has devoted a chapter in his *Construction of Reality in the Child* (1937). Only if the reader integrates these chapters with what Piaget presented in the first section of the book, does the fundamental relatedness of these concepts emerge. They are the constructivist substitute for 'categories' that Kant assumed to be a priori.

In order to maintain, for instance, that an object one experiences now is the self-same individual one experienced at some earlier point in the experiential flow, it becomes necessary to think of that object as perduring somewhere outside one's experiential field. This area, where objects might reside during the intervals in which one is not perceiving them, constitutes what I have called 'proto-space'. It is a space that has as yet no structure and no metric, and serves merely as a repository for objects that one can re-present to oneself but is not attending to at the moment. It is the space in which the child constructs an external world.

Similarly, once this proto-space has been conceived, the fact that one may have a succession of experiences while other objects are waiting there to be revisited by one's attention, leads to the construction of 'proto-time' as the medium of continuity that enables these waiting objects to conserve their individual identity. The continuum of proto-space and proto-time, then, constitutes the domain to which we refer when we use words such as 'existence' or 'being' in ordinary language. It should be clear that this domain is an abstraction from our experiential world and in no way entails the absolute ontology traditional philosophers want.

In the last chapter of *The Construction of Reality in the Child*, Piaget tackles the problem of the subject–object relation that has bedevilled western philosophy since its beginnings. Here, as in so many other passages of Piaget's writings, it is crucial to remember that he is concerned with *genetic* epistemology, i.e., with the ontogeny of knowledge, and not with ontology or the metaphysics of being. Thus he assumes a cognizing entity and proposes that this entity gradually distinguishes itself from all it is able to isolate and categorize as 'external' in its active experience. In a later work, he succinctly characterized this development:

> . . . at the termination of this [sensorimotor] period, i.e., when language and thought begin, he [the child] is for all practical purposes but one element or entity among others in a universe that he has gradually constructed himself and which hereafter he will experience as external to himself. (Piaget, 1967b, p.9)

The fact that Piaget continually ties his own construction of his abstract explanatory model of cognitive development to detailed observations and simple but often ingenious experiments with children, has apparently misled a great many readers (especially the more traditional psychologists) into focusing on empirical details rather than on the building blocks of the conceptual edifice he has created. The result has been a vast literature on Piaget's theory of cognitive development that largely disregards his epistemological presuppositions and consequently misinterprets the experiments as tests of performance rather than of conceptual operating.

Assimilation

Both assimilation and accommodation are key terms in Piaget's theory, and they are also among the most misunderstood. Assimilation is often described as 'the process whereby changing elements in the environment become incorporated into the structure of the organism' (Nash, 1970). This misleads, because it implies that the function of assimilation is to bring material *from the environment* into the organism. In my interpretation, assimilation must instead be understood as treating new material *as an instance of something known.* Piaget's own definition, can be found in many of his works. An example is the following:

> . . . no behaviour, even if it is new to the individual, constitutes an absolute beginning. It is always grafted onto previous schemes and therefore amounts to assimilating new elements to already constructed structures (innate, as reflexes are, or previously acquired). (Piaget, 1976a, p.17)

Cognitive assimilation comes about when a cognizing organism fits an experience into a conceptual structure it already has. A concrete, mechanical example of assimilation is what happens in those old-fashioned card sorting machines that work with punched cards. If one gives such a machine a deck of cards to compare to a model-card that has, say, three specific holes, it will pick out all the cards that have those holes, regardless of *any other* holes they might have. The machine is not intended to see these other holes and therefore considers all the cards it picks out as equivalent to the model-card. An observer, however, who does see the other holes, could say that the machine is assimilating all the cards to its model-card. The machine does not actively disregard other holes in the cards it examines, it simply does not perceive them.

Piaget borrowed the word 'assimilation' from biology. If someone eats an apple, one might say: His body is assimilating the apple. This does not mean that the apple is somehow modified to fit into the organism's structure, but it means that only certain chemical components of the apple are recognized as useful and extracted by the organism, while all others are ignored and thrown out. In the biological model, therefore, assimilation *does* take in elements of the environment — nutrients or other chemical substances. In the theory of cognition in which Piaget adopted the term, this is not so, because the operative processes are not physical transfer but perception and/or conception.

Once this is understood, the picture we get is quite different from the traditional one in which the senses are 'conveying information or data into the perceiving organism'. Using Piaget's definition, one can say: The cognitive organism perceives (assimilates) only what it can fit into the structures it already has. This, of course, is a description from the observer's point of view. It has actually the important implication that when an organism assimilates, it remains unaware of, or disregards, whatever does not fit into the conceptual structures it possesses.

Because no experiential situation in the life of an organism will be *exactly* the same as another, it is clear that in many cases it is advantageous (and therefore adaptive) to disregard differences. The peculiarity here, again from the observer's point of view, is that the adaptation seems to go in the opposite direction of the usual: perception modifies what is perceived in order to fit it into the organism's conceptual structures, whereas in the general biological sense, natural selection modifies the structure of organisms so that they fit within the constraints inherent in their environment. This apparent reversal of the adaptive process seems odd, as long as one thinks in terms of organisms perceiving objects that exist as such in an independent reality. From the constructivist point of view, however, adaptation does not mean adequation to an external world of existing things-in-themselves, but rather improving the organism's equilibrium, i.e., its fit, relative to experienced constraints. This is a crucial aspect of the constructivist model, and we shall return to it.

In short, assimilation always reduces new experiences to already existing sensorimotor or conceptual structures, and this inevitably raises the question why and how learning should ever take place. In this regard, too, Piaget has been widely misunderstood because many interpreters seem to have overlooked the fact that the Geneva school uses the terms assimilation and accommodation in the special context of what Piaget called 'schemes'.[9]

From Reflexes to Scheme Theory

Nowhere in Piaget's writings have I found a complete exposition of what I have come to call 'scheme theory'. However, indications that such a theory had, indeed, become a remarkably stable, unifying component of his thinking, can be found in most of his writings after 1935 (e.g., Piaget, 1937, 1945, 1967a). As Bärbel Inhelder remarked in her recent book: 'The notion of scheme has given and is still giving rise to different interpretations' (Inhelder and de Caprona, 1992, p.41). I certainly do not consider my interpretation the only, let alone the 'right' one, but it is the one we have found most useful, especially in analysing patterns of learning in mathematics and physics education.

Piaget's concept of the scheme also has its roots in biology. He was well acquainted with reflexes and investigated them in his own three children. Given that many reflexes or fixed-action patterns are fully operational in newborn infants, before any significant learning can have taken place, they must be considered wired in, i.e., genetically determined. In traditional biology textbooks, they are described as a concatenation of two things: a stimulus and a response, or a releaser and an action pattern.

$$\text{Stimulus} \longrightarrow \text{Response (Activity)}$$

Given that Piaget's interest had from the very beginning been focused on processes of adaptation, he saw very clearly that in order to become part of the general genetically determined characteristics of a species, these action patterns could be explained only as the result of natural selection. That is to say, organisms that manifested the reflexive action (because of accidental mutations) must have had a critical advantage over those that did not. Clearly this could not be due to the actions themselves but only to their result. He thus conceived of the reflex as consisting of three parts: A perceived situation, an activity associated with it, and a result of the activity which turned out to be beneficial for the actor.

The infant's rooting reflex, for instance, which makes the infant turn its head and search for something to suck whenever its cheek is touched, must have constituted a significant nutritional advantage. Individuals who did not have this automatic reaction, did not 'root' for the mother's breast, got an insufficient amount of milk, and were weeded out by natural selection.

Having constituted a three-part model of the reflex, it only needed the removal of the genetic fixedness in order to be applied to cognition. This, I am sure, was prompted by the simple observation that, at least in the higher mammals, most infantile fixed-action patterns are not as fixed as the biology textbooks would have it. In human animals, for example, the rooting reflex tends to disappear as their method of nutrition changes. Thus the reflex model could be adopted as explanatory tool in the domain of cognitively developed action and thought patterns that were in no way genetically determined. By viewing it from the organism's own point of view, it became the 'action scheme' and the basic principle of sensorimotor learning.

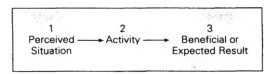

Figure 3.1:   Pattern of Action Scheme

This new perspective can be indicated by a change of terminology, and I have come to specify the three parts of schemes as follows:

1   Recognition of a certain situation;
2   a specific activity associated with that situation; and
3   the expectation that the activity produces a certain previously experienced result.

This tripartite pattern, I believe, is crucial for the proper understanding of the functioning of assimilation and accommodation.

The 'recognition' in part 1 is always the result of assimilation. An experiential situation is recognized as the starting-point of a scheme if it satisfies the conditions that have characterized it in the past. From an observer's point of view, it may manifest all sorts of differences relative to past situations that functioned as trigger, but the assimilating organism (e.g., the child) does not take these differences into account. If the experiential situation satisfies certain conditions, it triggers the associated activity.

Accommodation

The activity, part 2, then produces a result which the organism will attempt to assimilate to its expectation part 3. If the organism is unable to do this, there will be a perturbation (Piaget, 1974a, p.264). The perturbation, which may be either disappointment or surprise, may lead to all sorts of random reactions, but one among them seems particularly likely: if the initial situation 1 is still retrievable, it may now be reviewed, not as a compound triggering situation, but as a collection of sensory elements. This review may reveal characteristics that were disregarded by assimilation. If the unexpected outcome of the activity was disappointing, one or more of the newly noticed characteristics may effect a change in the recognition pattern and thus in the conditions that will trigger the activity in the future. Alternatively, if the unexpected outcome was pleasant or interesting, a new recognition pattern may be formed to include the new characteristic, and this will constitute a new scheme. In both cases there would be an act of learning and we would speak of an 'accommodation'. The same possibilities are opened, if the review reveals a difference in the performance of the activity, and this again could result in an accommodation.

Piaget's notion of scheme is not a simple affair. It cannot be properly understood unless one realizes that assimilation and accommodation are presumed to be subjective and depend on unobservable states in the particular cognizing agent. Assimilation has a generalizing effect in that it enables the agent to engage in a goal-directed action, even if, from the observer's point of view, the triggering situation is not quite the same as on previous occasions. If the goal is not achieved, the ensuing perturbation may lead to an accommodation. Either a new restrictive condition is added to the initial recognition process, which may serve in the future to prevent the particular 'unproductive' situation from triggering the activity. Or, if an unexpected result happens to be a desirable one, the added condition may serve to separate a new scheme from the old. In this case, the new condition will be central in the recognition pattern of the new scheme.

There is yet an added complication. The recognition of the activity's result 3 again depends on the particular pattern the agent has formed to recognize the results obtained in the course of prior experiences. That is to say, it, too, involves acts of assimilation.

Given this analysis, it is misleading to state, as do so many textbooks, that accommodation is simply the inverse of assimilation. In my interpretation of scheme theory, accommodation may take place only if a scheme does not yield the expected result. Hence it is largely determined by the cognizing agent's unobservable expectations, rather than by what an observer may call sensory 'input'.

In the context of accommodation, one might ask, what the situations are in which the child's schemes produce the perturbing outcomes that may impel it to learn. On the sensorimotor level, the permanent objects the child constructs, and the frequent interactions with them, continually provide innumerable opportunities to extend and refine the network of action patterns that constitute the 'physical' world. But the child's experiential world also comes to contain other people, and the almost constant interaction with them is an even richer source of perturbation and consequent accommodations. Piaget has stressed many times that the most frequent cause of accommodation is the interaction, and especially linguistic interaction, with others. Yet he is often criticized for not having taken into account the social component. The critics usually contend that adults or teachers transmit knowledge to children and students by interacting with them, and that certain forms of knowledge are inherent in society and transferred directly from the group to the individual. However, what mechanism could effect such a transfer from person to person, has never been explained. (That language cannot perform this service, was suggested in the preceding chapter by de Saussure's analysis and will be explained in greater detail in Chapter 7.)

The Concept of Equilibration

The notion of accommodation gives rise to an original theory of learning based on the concept of 'equilibration', a generic term for the elimination of perturbations. The focus on equilibration in his later work, is the reason why Piaget became interested in cybernetics (see Cellérier *et al.*, 1968; Piaget, 1977b).

Any control system that works with negative feedback has the purpose of eliminating perturbations in order to keep some chosen value constant (see Chapter 8). Consequently homeostasis became a central theme in control engineering. Piaget, however, made very clear from the outset, that what had to be kept constant in the cognitive context, did not have to be a fixed value, e.g., a set temperature in a thermostat, or the sugar level in human blood. It was more often a relation between changing values (as the equilibrium of a cyclist) or the regular change in some function.

Cognitive development is characterized by expanding equilibration (*équilibration majorante*), a term by which Piaget means an increase in the range of perturbations the organism is able to eliminate. One aspect of this notion of expanding equilibration is interesting for the philosophy of science and, I believe, also for the teaching of science. Every time the cognizing subject manages to eliminate a novel perturbation it is possible and sometimes probable that the accommodation that achieved this equilibration turns out to have introduced a concept or operation that proves incompatible with concepts or operations that were established earlier and proved viable in the elimination of other perturbations. When such an inconsistency surfaces, it will itself create a perturbation on a higher conceptual level, namely the level on which reflection reviews and compares available schemes.[10] The higher-level perturbation may then require a reconstruction on a lower level, before a satisfactory equilibrium can be restored.

The history of science shows many examples of this kind. At present, for instance, theoretical physicists are considerably perturbed by the fact that a model based on the concept of waves, works very well for the phenomenon of light under certain circumstances, but is incompatible with the corpuscular theory that seems to be required to explain the results of other experiments.

There is a further aspect of equilibration which, although not explicitly stated, is implicit in Piaget's repeated observation that the most frequent occasions for accommodation are provided by interactions with others. Insofar as these accommodations eliminate perturbations, they generate equilibrium not only among the conceptual structures of the individual, but also in the domain of social interaction. Had Piaget emphasized this implicit corollary a little more, the superficial criticism that his model disregards the social element would have been largely avoided.

As this brief exposition shows, scheme theory, like any other scientific account, involves certain presuppositions. According to it, cognizing organisms have to possess at least the following capabilities:

- The ability and, beyond it, the *tendency* to establish recurrences in the flow of experience;
- This, in turn, entails at least two further capabilities: remembering and retrieving (re-presenting) experiences, and the ability to make comparisons and judgments of similarity and difference; and
- The presupposition that the organism 'likes' certain experiences better than others; which is to say, it must have some elementary values.

These are features that clearly place Piaget's model in conflict with many twentieth-century psychologists, who diligently tried to avoid any reference to deliberate reflection, goal-directedness, and values.

Learning

The learning theory that emerges from Piaget's work can be summarized by saying that cognitive change and learning in a specific direction take place when a scheme, instead of producing the expected result, leads to perturbation, and perturbation, in turn, to an accommodation that maintains or re-establishes equilibrium.[11]

Learning and the knowledge it creates, thus, are explicitly instrumental. But here, again, it is crucial not to be rash or too simplistic in interpreting Piaget. His theory of cognition involves two kinds of 'viability' and therefore a twofold instrumentalism. On the sensorimotor level, viable action schemes are instrumental in helping organisms to achieve goals — sensory equilibrium and survival — in their interaction with the world they experience. On the level of reflective abstraction, however, operative schemes are instrumental in helping organisms achieve a relatively coherent conceptual network of structures that reflect the paths of acting as well as thinking, which, at their present point of experience, have turned out to be viable. The viability of concepts on this higher, more comprehensive level of abstraction is not measured by their practical value, but by their non-contradictory fit into the largest possible conceptual network. This aspect should put to rest the frequent complaint that constructivism undermines the practice of science. The first and essential criterion of viability on this second level is, indeed, analogous to what philosophers have called the 'coherence theory of truth', which concerns conceptual compatibility. Besides, just as in the case of scientific or philosophical models, other criteria, such as ease of handling, economy, simplicity, or what mathematicians call 'elegance', can be used to choose among models or theories that prove equally viable in the same set of circumstances.

The first kind of instrumentality might be called 'utilitarian' (the kind philosophers have traditionally scorned); the second, concerning conceptual coherence, is strictly epistemic and, as such, should be of some philosophical interest. It once more accentuates the radical shift in the conception of knowledge, a shift that eliminates the paradoxical conception of 'Truth' that requires the unattainable ontological test.

The step that substitutes viability in the experiential world for correspondence with ontological reality, applies to knowledge that results from inductive inferences and generalizations. It does not affect deductive inferences in logic and mathematics. In Piaget's view, the certainty of conclusions in these areas pertains to mental operations and not to the results of schemes on the sensorimotor level (see Beth and Piaget 1961; Glasersfeld, 1985).

With regard to conceptual learning, I want to stress a feature that is rarely discussed. Once experiential elements can be re-presented and combined to form hypothetical situations that have not actually been experienced, it becomes possible to generate thought experiments of all kinds. They may start with simple questions, such as: what would happen if I did this or that? And they may regard the most sophisticated abstract problems of physics and mathematics. Insofar as their results can be applied and lead to viable outcomes in practice, thought experiments constitute what is perhaps the most powerful learning procedure in the cognitive domain.

(*thought experiments*)

Different Types of Abstraction

Throughout Piaget's work the distinction he makes between 'figurative' and 'operative', and the concomitant distinction between (physical) 'acting' and (mental) 'operating', are indispensable for an understanding of his theoretical position.

'Figurative' refers to the domain of sensation and includes sensations generated by motion (kinaesthesia), by the metabolism of the organism (proprioception), and the composition of specific sensory data in perception. 'Acting' refers to actions on that sensorimotor level, and it is observable because it involves sensory objects and physical motion. Any abstraction of patterns composed of specific sensory and/or motor signals is what Piaget calls 'empirical'. The object-concepts the child constructs by associatively combining sensorimotor signals, are therefore 'empirical abstractions'.

In contrast, any result of conceptual construction that does not depend on specific sensory material but is determined by what the subject does, is 'operative' in Piaget's terminology. 'Operations', therefore, are always operations of the mind and, as such, not observable. Whatever results reflection upon these mental processes produces, are then called 'reflective abstraction'.[12] The material *from* which these abstractions are formed, consists of operations that the thinking subject itself performs and reflects upon. Hence there is an obvious analogy here to what Locke called the 'second source of ideas' (see Chapter 2).

There is one particular result of reflective abstraction that has been eminently fertile in the conceptual organization of our experiential world. Once a little reflection has recognized and isolated the basic procedure that in the past led to viable concepts of things and action schemes, this procedure can, as it were, be applied to itself. Put simply, this amounts to the following. The construction of object concepts and schemes is essentially *inductive*. By empirical abstraction, sensory particulars that recur in a number of experiential situations are retained and coordinated to form more or less stable patterns. These patterns are considered viable insofar as they serve to assimilate new experiences in a way that maintains or restores equilibrium. This simple form of the principle of induction, namely 'to retain what has functioned successfully in the past', can be abstracted and turned upon itself: because the inductive procedure has been a successful one, it may be advantageous to generate situations in which it *could* be employed. Consequently, a thinking subject that has reached this point by reflective abstraction and, for the time being, is not under pressure to cope with an actual problem, can imaginatively create material and generate reflective abstractions from it that may become useful in some future situation. This may involve material actually found in experience or it may take the form of a thought experiment with imaginary material.

Once the domain of the operative has been distinguished from that of the figurative, it becomes clear that the concept of object permanence is not an empirical but a reflective abstraction, because it does not derive from sensorimotor material but from the operative conceptual construct of individual identity.

The fact that this distinction was not, or not fully understood, underlies much of the controversy about the notorious Piagetian tasks. It has, for instance, been demonstrated innumerable times that not only a child but also a good many animals can be trained to react to the appearance of a given object in the same way they reacted to it before it disappeared. This indicates that the subject has formed a specific object concept (a figurative achievement); but it in no way proves, as is often claimed, that it constitutes a demonstration of the concept of object permanence (which is an operative achievement). To justify that second claim, it would be necessary to show (1) that the organism *believes* that the object in question 'exists' somewhere while it is not being experienced, and (2) that the organism is able to produce a re-presentation (i.e., a visualized image) of the object when it is not within the organism's actual perceptual field.[13]

The fundamental difference between the conventional psychologist's empirical approach and Piaget's is that the first focuses on observable behaviour and performance, whereas the second focuses on the results of reflective abstractions, that is, mental operations. Since these operations are never directly observable, they can only be *inferred* from observation. As a rule, such inferences are not possible from a single observation but require a sequence stretched over time.

Any reference to mental operations does, of course, raise a formidable problem — the problem of consciousness. In Piaget's theory it crops up in various places, because the four capabilities I listed as presuppositions would seem to require consciousness, at least at the higher stages of cognitive development. In Piaget's model, certain operations are sometimes said to be manifestly under the control of consciousness, at other times not. He has clearly shown this experimentally in several of his books (e.g., Piaget, 1974a, 1974b, 1977a), but the phenomenon of consciousness remains mysterious. He himself said:

> Psychology is not a science of consciousness, it is a science of behaviour! One studies behaviour, including the *attainment* of consciousness where one can get hold of it, but where one cannot, it is not a problem. (Bringuier, 1977, p.180, my emphasis)

In this he does not differ from other researchers. I know of no one today, who has produced a viable model of consciousness, yet we are mostly able to make reliable inferences as to when a human actor is conscious and when not.

Stages of Development

Piaget's theory has been correctly described as a 'stage theory' because it segments the cognitive development from birth to maturity into successive stages. In this regard, too, there are widespread misunderstandings. Given his genetic epistemology and its denial of the traditional notion that knowledge should be a picture of reality, it follows that, from his point of view, whatever theory a psychological investigator builds up, it will not be a description of the observed subjects' *objective* mental reality but rather a conceptual tool for systematizing the investigator's experiences with the subjects. All observation is necessarily structured by the observer and, as Piaget said:

> I think that all structures are constructed and that the fundamental feature is the course of this construction: I think that nothing is given at the start, except some limiting points on which all the rest is based. The structures are neither given in advance in the human mind nor in the external world, as we perceive or organise it. (ibid., p.63)

The stage theory, therefore, should be taken for what it is, namely a more or less successful way of organizing an observer's view of developing children.

Apart from this, it must be said that Piaget changed his view, or rather his assessment of the importance of stages. At the beginning, he was inclined to believe that once an operation characterizing the next higher stage was manifested by the child's behaviour, this operation would be available to the child in all contexts where it might be relevant. This turned out not to be the case. The use of any given mental operation is now considered far more context-dependent than was originally assumed. Hence it can take a certain time for a particular way of operating to spread to other contexts (horizontal *décalage*). This means that a child may, for instance, have demonstrated 'formal operating' in a given context, while in other contexts it is still in a preceding stage. What has remained intact, however, is the assumption that there is a relatively fixed order in the acquisition of the operations that characterize the stages (vertical *décalage*).

The Observer and the Observed

One of the presuppositions of Piaget's theory is that the thinking subject has two basic capabilities. First, it can coordinate elements of sensory and motor experience; second, if the conceptual structures resulting from such coordination turn out to be viable in further experiential situations, it is able to abstract, from its own operating, regularities and rules that may help to manage future experience. The elements the thinking subject coordinates are by definition present *in* the subject's system because they are 'experiential'. The system has no access to items which, from an observer's point of view, are seen as external, 'environmental' causes of the system's experiences. Coordination, thus, is a strictly internal affair and, therefore, it is always subjective to the coordinator. This applies not only to the children Piaget was studying, but also to every observer, be it a scientist or a simple bystander. No one can escape this fundamental subjectivity of experience, and the philosophers who purport to have access to a 'God's eye view' are no exception.[14] Like all cognizing organisms, they draw conclusions from their own sensorimotor and conceptual experience, and any explanation of their conclusions, i.e., their 'knowledge', must be in terms of *internal* events and cannot draw on elements posited elsewhere.

Piaget made a clear distinction between the points of view, on the one hand, of the living, experiencing subject itself and, on the other, of the observer who is trying to understand how such a subject can construct knowledge.

> In the first place, one has to distinguish the individual subject, . . . and the epistemic subject or cognitive core that is common to all subjects at the same level. In the second place, one must contrast, on the one hand, the attainment of consciousness (which is always fragmentary and often distorting), and on the other, what the subject succeeds in *doing* in its intellectual activities of which it knows the results but not the mechanisms. But in dissociating the subject from the 'self' and what it 'lives', there remain its *operations*, that is to say, what it draws by reflective abstraction from the generalised coordinations of its actions. (Piaget, 1970b, p.120)

It is the observers who, in order to construct a model of cognition, 'dissociate' from the observed living subject what they categorize as coordinations and the results the subject draws from them by reflective abstraction. Only observers can refer to a subject's interaction with its environment and characterize the relation between the subject's structures (biological as well as conceptual) and the world in which, from an observer's point of view, the observed subject lives and operates.

Experience and Reality

In Piaget's model, then, the subject's interactive contacts with its environment are always and necessarily of the same kind: a conceptual structure fails because it does not lead to the result the subject has come to expect of it. Cognitive structures, it must be remembered, are tied to action and to use. Action and use are something more than random motion or random change — they take place in the context of 'action schemes'. This radically differentiates Piaget's approach from both the behaviourist's stimulus–response approach and the physicist's linear cause–effect chains, because action schemes are explicitly goal-directed. As Piaget himself has occasionally suggested, action schemes are rather like feedback loops, because their inherent dual mechanisms of assimilation and accommodation make them self-regulating and therefore circular in that specific sense (the cybernetic parallel will be treated in Chapter 8).

The relation of knowledge to the real world, thus, is reciprocal in Piaget's model, because any conceptual structure is likely to be modified when it clashes with a constraint. To the thinking subject, the environment could manifest itself only through such clashes. The subject, therefore, can *know* no more than that certain structures and schemes have clashed with constraints, while others constitute a viable way of managing.

This is analogous to saying that the biological organisms that are alive at a given moment, are viable because they have so far managed to survive. To infer from this relation a likeness or 'correspondence' would be a *non sequitur* and a gross misrepresentation. Having avoided clashes with obstacles does not tell us what the obstacles *are* and how a reality consisting of them might be structured. The experience of a clash or failure merely tells us that, under the particular experiential circumstances, the particular scheme did not work. The failure, moreover, may be due, not to the world but to a snag or contradiction intrinsic to the scheme. If, instead, a scheme is successful, this merely shows it to be viable in that it 'worked'. No inference about a 'real' world can be drawn from this viability, because a countless number of other schemes might have worked as well.

The most important consequence of this model of cognition can be summarized as follows. Our knowledge of clashes with what we have categorized as 'environment' or 'real world' can be articulated and re-presented only in terms of viable conceptual structures, i.e., structures which, themselves, have *not* come into contact with obstacles. At best, then, this knowledge of clashes and failures describes reality in 'negative' terms. Any notion that cognitive structures could come to reflect ontological reality — e.g., that we could discover the ontic shape of things by sliding our senses or measuring instruments along the surfaces of things-in-themselves and thus plot deliberate contacts — is an illusion. The space and time in which we move, measure and, above all, in which we map our movements and operations, are our own construction, and no explanation that relies on them can transcend our experiential world.

In short, the epistemological view which I find to be the most compatible with Piaget's work is an instrumentalist one in which knowledge does not mean knowledge of an experiencer-independent world. From this perspective, cognitive structures — action schemes. concepts, rules, theories, and laws — are evaluated primarily by the criterion of success, and success must ultimately be understood in terms of the organisms' efforts to gain, maintain, and extend its internal equilibrium in the face of perturbations.

Conclusion

This attempt to lay out some of the key concepts in Piaget's model of cognition and cognitive development is far from complete. I have tried to focus on those points which, it seems to me, are most important yet frequently misrepresented and misunderstood.

As I suggested at the beginning, Piaget's writings do contain contradictions. But if one searches his work for all that can be incorporated in a consistent model of human knowing, one does, I believe, come to the conclusion that the occasional passages that imply a realist stance are nothing but slips of mind. In all his pioneering he may every now and then have lapsed into the ordinary, current ways of speaking that belong to the traditional epistemology he was struggling to overcome.

Notes

1 The official Piaget bibliography (Archives Jean Piaget, 1989) lists a total of 1232 titles, including revised editions and translations.
2 It has been suggested that Piaget's approach was anticipated by James Mark Baldwin, but Vonèche (1982) has shown that the connections are tenuous and that the two authors developed most of their ideas independently.
3 Rita Vuyk, whose two volumes of *Overview and Critique of Piaget's Genetic Epistemology* (1981) are among the best attempts, makes a remark in her preface that I would adopt unconditionally: 'All the passages annotated in my copies of his books as being incomprehensible, nonsense, contradictory, etc., have been deleted from the overview' (p.ix).
4 I owe this quotation to Rita Vuyk, who used it as motto in her *Overview* (1981, p.v).

5   The Baccalaureate is conferred at the successful termination of high school.

6   It is important to realize that the neural network is constantly teeming with signals that originate in the peripheral neurones that are usually called 'receptors' or 'sensory organs'. While you are reading this, there are innumerable signals available to you to which you are not attending; e.g., some that you would call 'tactual' that originate in your rump and which you could interpret as telling you that you are sitting; others that originate in your ears and which you could interpret as telling you that a car is passing in the street; but your attention was focused on this text and therefore you were not doing any of this other interpreting before I mentioned the possibility. Similarly, literally millions of signals are constantly generated in the retinas of your eyes, but you disregard almost all of them because you are focusing your attention on 'some specific thing', i.e., a coordination of signals that is of interest and 'makes sense' to you at the moment.

7   Psychologists might call this 'constancy', but I prefer the somewhat archaic word 'perdurance' to distinguish the purely conceptual notion from the perceptual 'constancies'.

8   This, clearly, is a conceptual affair and not a matter of observable actions. It can be ascertained in the course of interaction, but not by simply recording behavioural responses. This has frequently been misunderstood by developmental psychologists who were looking for observable manifestations of 'object permanence' in children, and by animal psychologists who attempted to demonstrate it in animals.

9   The French word is *schème* and it refers to a scheme of actions or operations. Unfortunately it has often been translated as 'schema' (plural 'schemata') which corresponds to the French word *schéma*, a word rarely found in Piaget's texts because it refers to static diagrams such as town plans or flow charts. By disregarding the difference, translators have caused considerable confusion among English readers.

10 Note that this reflective review of available schemes anticipated and included the capability for which contemporary psychologists have invented the term 'metacognition'.

11 Needless to say, there is also a great deal of accidental learning that arises from the perturbations generated by unexpected sequences of experience.

12 Piaget divided this form of abstraction into four categories (1977); they will be discussed in Chapter 5.

13 With 2-year-old children (and sometimes younger ones) there is an intermediate indication that the concept of object permanence is about to be achieved: the infant, not having found the object in three of the four known hiding places, shows by expression and body language that it *knows* the object will be in the last one.

14 This expression is Hilary Putnam's (1981).

Chapter 4

The Construction of Concepts[1]

As I described at the beginning of this book, my interest in theories of knowledge was triggered by the use of different languages and the early discovery that each was tied to a different experiential world. At the same time it seemed that they all functioned in much the same way, and I began to look for a model for the stuff that we call 'meaning'.

Sensorimotor knowledge manifests itself in actions, but conceptual knowledge is expressed in symbols. When we come to investigate this knowledge, the symbols are mostly linguistic. Therefore, semantic analysis, i.e., the analysis of meaning, has to be an important facet of any theory of knowing.

The relation between conceptual structures and their linguistic expressions was also at the heart of the Italian Operationist School, and Ceccato's method for the analysis of meaning came to play an important role in the development of the constructivist theory. I called it 'conceptual semantics' and continued to use it during my work on machine translation. It is an unconventional method and differs sharply from the common practice in linguistics. It does not try to find appropriate verbal definitions of words, as one might find in a dictionary, but instead, aims at providing 'recipes' that specify the mental operations that are required to obtain a particular concept. It was a sophisticated application of Bridgman's idea of operational definition. One might be tempted to speak of an analysis of mental behaviour but, given current usage, this would be counterproductive.

In the United States, where I have been living for the last quarter of a century, psychology has chosen to define itself as the science of behaviour — and behaviour, as the followers of Watson and Skinner preached with devastating success, is what we can *observe* an organism do. The founders of behaviourism were adamant in their contention that there is nothing beyond the observable that could be of interest to science.[2] Focusing exclusively on behaviour and defining behaviour as observable responses, makes it easy to avoid dealing with any intelligent organism's more complex capabilities. In the long run, it provides merely partial models of the behaviour of pigeons and rats.

Piaget, too, described psychology as the science of behaviour (see Chapter 3), but it was for a different reason. Behaviour was important to Piaget, because an observer can often infer from it what might be going on in another person's mind, and the functioning of the mind was his primary interest.

Among the most intriguing human activities that can never be directly observed is thinking or reflecting. At times one can infer thoughts or reflections from a facial expression or a position — as Rodin hoped when he moulded his *Penseur* — and sometimes one might infer them from subsequent acts of speech or other actions. But the actual process of thinking remains invisible and so do the concepts it uses and the raw material of which they are composed.

Adult human beings, however, usually speak some language, and this entails that most, if not all, their concepts have to be associated with words. This opens a window on conceptual structures. Speakers of a language can examine concepts they habitually employ in thinking. Not by direct introspection, but by imagining a variety of related situations and asking themselves: 'what word would fit, if I wanted to describe this particular part or aspect of my experience'.

William James, whose powerful analytic mind often focused on problems of meaning, told a delightful story that provides a vivid illustration of conceptual analysis:

Some years ago, being with a camping party in the mountains, I returned from a solitary ramble to find every one engaged in a ferocious metaphysical dispute. The *corpus* of the dispute was a squirrel — a live squirrel supposed to be clinging to one side of a tree-trunk; while over against the tree's opposite side a human being was imagined to stand. This human witness tries to get sight of the squirrel by moving rapidly round the tree, but no matter how fast he goes, the squirrel moves as fast in the opposite direction and always keeps the tree between himself and the man, so that never a glimpse of him is caught.[3] The resultant metaphysical problem now is this: *Does the man go round the squirrel or not?* He goes round the tree, sure enough, and the squirrel is on the tree; but does he go round the squirrel? In the unlimited leisure of the wilderness, discussion had been worn threadbare. Everyone had taken sides, and was obstinate; and the numbers on both sides were even. Each side, when I appeared, therefore appealed to me to make it a majority. Mindful of the scholastic adage that whenever you meet a contradiction you must make a distinction, I immediately sought and found one, as follows: 'Which party is right', I said, 'depends on what you *practically mean* by "going round" the squirrel. If you mean passing from the north of him to the east, then to the south, then to the west, and then to the north of him again, obviously the man does go round him, for he occupies these successive positions. But if on the contrary you mean being first in front of him, then on the right of him, then behind him, then on his left, and finally in front again, it is quite as obvious that the man fails to go round him, for by the compensating movements the squirrel makes, he keeps his belly turned towards the man all the time, and his back turned away. Make the distinction, and there is no occasion for any farther dispute. You are both right and both wrong according as you conceive the verb "to go round" in one practical fashion or the other.' (James, 1907/1955, pp.41–2)

The analysis of the experiential situation is 'logical' and it brings to the surface a difference of conceptualization that is blurred by the ordinary use of language. The possibility of bringing such distinctions to awareness by examining the meaning of words, was the starting point that Silvio Ceccato's group chose for their conceptual analyses in the 1940s. But Ceccato had added a second question that led the enterprise beyond language and into the very domain of thought: 'what mental operations must be carried out to see the presented situation in the particular way one is seeing it'. This was the first serious application of the method the Nobel laureate Bridgman (1936) had called 'operational definition'. When this method was first published, it created quite a stir in Bridgman's field of theoretical physics. A decade or so later, the psychological establishment picked up the part of Bridgman's idea that focused on physical operations. All that concerned abstractions was suppressed. As a result, we have the appalling slogan, 'Intelligence is what the intelligence test measures', a slogan that caused innumerable gifted children all over the world to be hopelessly misjudged.

Bridgman had something more sophisticated in mind. Speaking of mathematics, he explicitly said that we can also construct concepts in other ways and then experiment with them

. . . to see whether the concepts are useful. We still have operational meaning for our concepts, but the operations are mental operations, and have no necessary physical validity . . . But even mental operations are subject to certain limitations, and if we transgress these in formulating our tentative concepts we may expect trouble. In particular all mental operations must be made in time, and are therefore ordered in time. (Bridgman, 1936/1964, p.11)

In Bridgman's work one finds examples of physical operations, such as measuring the distance between two points by successively placing a meter on the line that connects them. Of the mental operations he gives some results, but he does not specify how he imagined them to work. This is the point that Ceccato pursued. He came up with a model of a mental procedure based on the notion of a pulsating attention and the ability to form combinatorial patterns of such pulses. (This will be explained in more detail in the section on the conception of number, in Chapter 9).

For the somewhat grosser level of analysis used for the non-mathematical concepts I want to present here, it will be sufficient to say that Ceccato's method consisted in viewing sensory experience much like a cinema film, made up of a sequence of still 'frames' that give rise to concepts of change, movement, extension, etc., when they are presented in rapid continuous succession.[4]

Analysis of Operations

In the preceding chapter, mental operations were shown to be an integral part of Piaget's theory. They are the basic element of reflective abstractions and of everything Piaget's theory subsumes under the term 'operative'. Thus we now have three authors who speak of mental operations and, of course, there are differences. Bridgman must be acknowledged for establishing the thinking subject's non-physical operations as a respectable topic for investigation. The distinction was the important feature for him, rather than any functional details of the mind.

For Piaget, mental operations are either 'abstraction' ('empirical', when abstraction is made from sensorimotor material, 'reflective', when it concerns the subject's own activity) or 'logical transformation' (changing the relations that characterize a group structure). In contrast to both, Ceccato's idea of successive pulses of attention provides a mechanism, a hypothetical model of *how* the mind might operate. Although the three conceptions are not interchangeable they overlap insofar as they refer to the mind's activity, and are mutually compatible in my interpretation. The authors are speaking about different levels of analysis. Bridgman classified concepts as he found them; Piaget examined their logical character as elements of cognition; and Ceccato analysed them from the perspective of a technician who intends to build a functioning model of the adult mind.

As far as one can tell, Bridgman had no contact whatever with the two other authors. Piaget was well aware of Bridgman's work and explicitly referred to it (see Piaget, 1957, p.7). He also met Ceccato at least once and included him in the original editorial committee of the *Études d'épistémologie génétique*, a chain of thirty-seven publications that continued until 1980. However, given the lack of a common language and an apparent clash of personalities, they never came around to discussing the fundamental ideas they could have agreed on.

In my view, it is something of a tragedy that Piaget and Ceccato could not work together. Ceccato's model of the functioning of attention is just the sort of thing that is needed to underpin Piaget's crucial notion of abstraction. It might have opened the path to neurophysiological experimentation which would have provided a welcome connection between Piaget's theoretical model and empirical research. But this is not what I want to discuss here. Rather, I want to present some results of conceptual analysis by means of Ceccato's model of successive frames.

The Concept of Change

One of the conceptual structures that plays a major role in providing a fit with our experiential world, is the concept referred to by the word 'change'. Though we cannot watch how a language-user builds up his or her concepts, we can investigate them by doing two things. First, examine what kind of situations the word is intended to describe; second, try to unravel, from a logical point of view, what elements the associated concept must incorporate in order adequately to reflect certain experiential situations. If we do this with the concept of change, we can say straightaway that we would not have occasion to conceive of change if we had no memory. In order to speak of change, we have to consider at least two moments of experience and spot a difference.

This need of more than one experience was implicit already in Zeno's paradox of the arrow. If you watch an arrow flying through the air, you see it move and change place from the moment it leaves the bow to the moment it hits the target. If, however, you consider it at any one moment during its flight, it does not move. Zeno knew nothing of movies, but what we today have in the form of cinema film is a perfect illustration of what he was suggesting. A film showing the flying arrow would be made up of a series of still frames. Each frame would show a stationary arrow at a slightly different place. If we saw only a single frame, we might guess that the arrow was moving, but this would be an inference made by analogy to other experiences we have had of arrows. The single frame itself contains no movement.

This is no new revelation. Jeremy Bentham stated it clearly:

> When of any body it is said: 'That body has been in motion', what is meant is that, at or in different portions of the field of time, that body has occupied different portions in the field of space. (Bentham, in Ogden, 1959, p.115)

Knowledge of the movement must be constructed by the observer in his or her field of experience. Notice that I am concerned with *knowledge* of the movement, not with the question of whether or not the 'real' arrow moves. The cinema film is a good illustration, precisely because on it the arrow does *not* move. Yet when we see the film projected, we see the movement. Thus the question arises: how can this experience be generated?

The analysis must begin with the fact that we need at least two consecutive experiential frames. Ceccato's method consists in mapping the minimum requirements for each frame. We therefore mark two moments of the experiential flow: t_1 and t_2.

To speak of 'change', we also need the perception or conception of a difference. For example, a difference of colour, shape, size, location, or the like. If the frames contain a background, the arrow's location can be defined relative to some other visible item; if nothing else is shown, it can be defined only in relation to the edges of the frame. If the two frames showed the arrow in the same place, you could not say that it was moving. A difference in location by itself, however, is not enough — it has to be attributed to some *thing* of which, as in the case of the arrow, we can then say that it has moved.

If, instead of an arrow, I showed you a small green plum and then another larger purple one, you would not be inclined to speak of change. But if the green plum were on a tree, and a few weeks later you looked at it again and saw the same plum purple, you might say that its colour has changed or, indeed, that it has ripened. In other words, the concept of 'change' requires a difference perceived in an object that is considered the same object at two moments in the flow of experience.

But sameness, as I explained in the context of 'object permanence', is not as simple a notion as it might seem.

William James anticipated the crucial distinction to be made:

> Permanent 'things' again; the 'same' thing and its various 'appearances' and 'alterations'; the different 'kinds' of thing . . . it is only the smallest part of his experience's flux that anyone actually does straighten out by applying to it these conceptual instruments. Out of them all our lowest ancestors probably used only, and then most vaguely and inaccurately, the notion of 'the same again'. But even then if you had asked them whether the same were a 'thing' that had endured throughout the unseen interval, they would probably have been at a loss, and would have said that they had never asked that question, or considered matters in that light. (James, 1907/1955, p.119)

Many of our contemporaries have not considered that question either. The sameness that you might ascribe to the wine glasses on a dinner table or to the chairs around it, would be the sameness of 'equivalence' and it would not lead you to speak of change. The sameness involved in the construction of the concept of change has to be 'individual identity'. That is to say, the thing about which we want to say it has changed, must be the self-same individual in both frames.

I can now complete the schematic representation of the concept, indicating the arrow by 'X' and marking its different locations. The fact that the X refers to the self-same arrow in subsequent frames (individual identity) is shown by an equal sign with three, instead of two lines.

$$
\begin{array}{ccc}
t_1 & & t_2 \\[4pt]
X & \equiv & X \\
\text{at} & & \text{at} \\
\text{location A} & \neq & \text{location B}
\end{array}
$$

Figure 4.1: Change of Location

In the example of the plum, the structure of the change-diagram would be similar, but instead of locations, which determine change in the arrow experience, different colours would be associated with X at t_1 and t_2.

The Concept of Motion

The case of motion, however, is more complicated. Not because the diagram does not represent the core of the concept adequately, but because, on the one hand, it does not specify what kind of motion one has in mind, and on the other, the words that refer to the concept are often used loosely. The diagram, for instance, shows that the arrow moved, but it does not specify whether it did so because it was shot from a bow or because the archer carried it in his quiver from location A to location B. Thus, when we used this method for the analysis of specific pieces of 'language' or text, it had to be expanded to include conventional ways of indicating contextual features. What I called 'looseness', however, concerns the use of concepts and frequently does reflect features that are relevant to conceptual analysis.

Let us assume that one of the situations to be examined in the course of the investigation of the concept of motion, would be described by the phrase 'This bus goes from Exeter to London'. It could at once be represented by the above diagram, because the locations A and B can easily be specified respectively as Exeter and London, and X could stand for the bus. But what if the description were 'This *road* goes from Exeter to London.'? Now it does not seem to involve any motion. You might shrug your shoulders and conclude once again that the way language is used is not a logical affair.

As it happened, this example, and others like it, played an important role in the conceptual analyses we carried out in the 1960s and, as far as I am concerned, it helped in the development of the constructivist model of cognition. The attempt to explain it highlights a crucial aspect of mental operations.

In order to understand the 'meaning' of the phrase 'This road goes to London', you have to bring forth, as in the first case, an image (i.e., a re-presentation) which, at the moment, is your interpretation of the word 'London'. It may be your image of Trafalgar Square, a particular London pub, or the bus terminal. Whatever it happens to be, it will be a location different from the location of Exeter. The expression 'goes to' still indicates a change of place but no likely agent of actual physical motion is specified. Yet, the concept of 'road' implies unlimited extension in one dimension, the spatial connection of locations, and the possibility of travel. This is sufficient for the reader to transform the displacement of attention from 'Exeter' to 'London' into the potential motion of a physical object.

Linguists and philosophers of language have long come around to the idea that the meaning of words depends to a large extent on the context in which they are found. In fact, a word, as Roland Barthes said of literature in general: 'sets up ambiguities, not *a* meaning' (Barthes, 1987, p.72). There is hardly ever a simple one-to-one relation between a word and a concept. Prepositions are a good example (see the different uses of the English 'by' listed in Chapter 1). One way to think of this relative looseness is to think of mental operations as tools that serve to establish basic relations between elements of experience. The basic relations are then further specified by the given context. One could say, for example, that 'by' designates, among other things, the relation of *closeness* in space or time and that it is the context that determines in instances such as 'by the river' and 'by Friday' whether the closeness is to be visualized as proximity or coincidence in space or time; and in instances such as 'by force' and 'by moonlight', as the more specific relation of means and ends. In principle, this is similar to the case of the bus and the road: the one implies actual change of place, the other potential travel.

The different interpretations that contexts can elicit for a word, however, are frequently not all the same for the word of another language, even if that word seems to be similar in many respects. A little story may illustrate this. One Sunday, I was walking with an English friend along a river in the countryside near Milan. For a short stretch railway tracks followed the river, and we had to cross them into a field. There a family, complete with *mamma* and *bambini* had spread out a picnic and the children were running about in the grass. Suddenly a distant rumbling noise could be heard, and the mother jumped up and shouted: '*Attenti bambini — arriva il treno!*' My friend asked what she had said, and I was about to translate literally: 'Be careful children, the train is arriving', but I realized that in English she would not have said that the train was arriving, but that it was *coming*.

When I came home, I drew up the diagrams for the situation. For the English verb 'to arrive' one needs at least three frames. Two to indicate that the active item X changes location, and two to indicate that it comes to a state of rest.

t_1		t_2		t_3
X	$\equiv$	X	$\equiv$	X
loc. n	$\neq$	loc. 'here'	$=$	loc. 'here'
	(motion)		(state)	

Figure 4.2: Diagram of the English Verb 'to Arrive'

The change from location 'n' (anywhere) at t_1 to location 'here' at t_2, indicates *motion*; the location remaining the same in t_2 and t_3, indicates *state*.

The Italian verb *arrivare*, although it derives from the same Latin root, indicates a process of motion but not a necessary stop. Hence it would be mapped as:

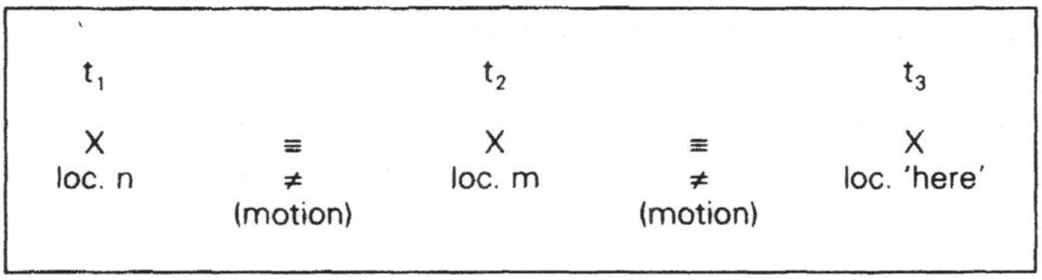

t_1		t_2		t_3
X	≡	X	≡	X
loc. n	≠	loc. m	≠	loc. 'here'
	(motion)		(motion)	

Figure 4.3: Diagram of the Italian Verb 'Arrivare'

In most situations, this conceptual difference would not be relevant for translations from English to Italian and vice versa, because the expectation that the item X comes to a stop would not be incompatible with the representation the text intends to evoke. Indeed, small bilingual dictionaries of the two languages give 'to arrive' and *arrivare* as synonyms without further explanation.

The analysis of conceptual structures by means of comparisons of their expression in different languages proved a laborious but fertile undertaking. It brought out a great many subtle distinctions, which native speakers make in their own languages without becoming conscious of them.

Generating Individual Identity

The method of graphic mapping may help to make clearer the verbal description I gave earlier of the concept of 'object permanence'. I explained that Piaget's *La construction du réel chez l'enfant* contains chapters on the construction of the concepts of object, space, causality, and time, and that he saw these conceptual developments as happening simultaneously in the child's mind, arising interdependently out of the same basic material. At least some of this raw material can be isolated and represented in a diagram.

The key to 'object permanence' is the constitution of individual identity. This means that two experiences of an object are linked by means of the idea that the object has remained one and the same. Presumably it derives from very early experiences when the infant's attention is caught by, and remains on, a moving object in its visual field. It may be a person that walks through the room or a toy that is being moved by someone. Whatever the object, the child keeps it in focus and moves its head, or at least its eyes, to follow. Psychologists have called this 'visual tracking'. When the object is continuously in view, there is no question that it remains one and the same, simply because it is continuously present.

However, experiments have shown that if the object moves at a regular pace and disappears behind something that screens it from the child's view, the child will continue the tracking movement, and its gaze will be right there at the other end of the screen when the object reappears (see Bower, 1974, p.195ff). Thus, a connection between the experiences of the object before and after its disappearance, is supplied by the observer's own movement. Much later, when concepts are abstracted from experience, this continuity of a tracking movement does become a powerful reason for preserving the individual identity of the tracked object in spite of an interval during which it moves out of sight. Of course, there are also other criteria, such as particular colour or shape, or marks that are considered individual characteristics. In the case of human individuals, there are birth marks, scars, dental fillings, and finger prints. But all these indications of identity are, as it were, surface criteria, and where persons are concerned, we have a much deeper one that overrides all others and is considered decisive by itself.

Imagine that your brother, who is two or three years older than you, left the family and went to Australia when you were still in your teens. No one you know has seen him or heard from him since. Now that you are middle-aged, a man comes to visit you and claims to be your brother. He does not fit your remembered image. He has lost his hair, wears glasses, and speaks with a wholly unfamiliar accent. In short, his appearance is that of a total stranger. Is he an impostor? But then he says: 'Do you remember the time we broke the Jones' garage window and blamed it on the Irish kids down the street?' — Maybe you had repressed it, but now you remember the shameful act. It instantly convinces you that the visitor *is* your brother, because no one else knew of that incident, and you consider it highly unlikely that your brother would have boasted about it to others.

It seems that memories are the most reliable indicator of individual identity. We firmly believe that they are the most individual personal possession and unique — especially if they are embarrassing. It has struck me that this may one of the reasons why we have a profound reluctance seriously to consider the possibility of telepathy. For if we believed that our thoughts are indiscriminately accessible to others, the most reliable criterion of individual identity would be put in question because it is dependent on the absolute privacy of our thoughts.

In any case, the concept of individual identity can be mapped by a diagram. It is the posited identity of an experiential object that, during one or more attentional frames, is not present in the subject's perceptual field. It consists of the following sequence:

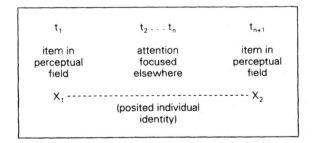

Figure 4.4: The Generation of Individual Identity

At t_1 the child isolates an item in its perceptual field; at times $t_{2...n}$ the item is no longer perceived and the child's attention is focused elsewhere; at time t_{n+1} the child again isolates an item X_2 in its perceptual field and considers this second item the self-same individual item as at time t_1.

Note that if the two items are not considered to be the self-same individual, but merely equivalent, the sequential structure becomes the model of classification. The difference between the two conceptual constructions lies in the kind of sameness used to link the two experiences: 'individual identity' in one, 'equivalence' in the other.

In both cases, the conceptual structures require a further element for their completion: The child must be able to visualize the object when it is *not* in the perceptual field. This ability yields what is properly called a 're-presentation', which is in fact a re-play or re-construction of a past experience. This is the point that was overlooked by many critics of Piaget and by all the experimenters who tried to demonstrate object permanence in rats, cats, and rhesus monkeys. They did not and, I believe, could not, show that the animal had a re-presentation of the object in question.

Space and Time

One of the perspectives opened by the structural pattern of change is the following. On the one hand, the co-involvement of both the notions of sameness and difference creates the appearance of contradiction and can, therefore, generate a perturbation. On the other hand, if the first notion of causality arises, as Piaget believed, from the reiteration of actions that lead to an interesting result, it seems plausible that a novel property in an object that is nevertheless considered the self-same individual would stir some interest. This interest, at some later stage, would lead backwards to the situation at t_1 — the remembered situation prior to the interesting result — in order to discern something that could be held causally responsible for the object's novel property in the later frame. Such an exploration would be the beginning of research, and it would quickly lead to the strategy that underlies the kind of scientific experiments that aim at establishing the cause of a given phenomenon.

A second perspective is opened by the fact that there are attentional frames in which the object that later is considered one-and-the-same, is not present in the actual experiential field. This raises the question where it might be and what it might be doing while one's attention is focused on other things. As I suggested earlier, this requires the conception of a kind of resting place where objects can maintain their identity without being perceived. I called this resting place a proto-space because, at first, it has no articulation or relational structure. Initially, it is an amorphous repository, but as it is furnished with permanent objects which are related by visual and physical movements whenever they are actually in the experiential field, the relations abstracted from these movements provide the repository with the permanent structure of more-dimensional space. A detailed account of this construction has been given by Poincaré (1952, p.51*ff*).

The fact that the object is considered the self-same individual in attentional frames that are not consecutive, requires that its identity be stretched across an interval. This means that a continuity has to be constructed that lies *outside* the experiential field, a continuity that links the present experience of the object to an experience in the past. This outside continuity that runs parallel to a sequence of actual experiences, I have called proto-time. It, too, is at first an undifferentiated, eventless continuity. But then the sequence of the subject's actual experiences that were lived during the interval, can be mapped onto it and thus give it a vicarious segmentation. Whenever this happens, when a sequence of actual experiences is projected on the eventless continuity of permanent objects in the repository, the concept of 'time' can be created. The structure of temporal relations is thus generated by superimposing a sequence of actual experiences on a continuity that is not in the experiential field and has itself no articulation. This gives these experiences duration and provides the directionality that it is sometimes called 'time's arrow'. Finally, when the projected sequence of experiences is a regular one, such as night and day, the seasons, or the movement of stars, the segmentation of the continuity becomes uniform and thus turns the amorphous proto-time into real time that can be measured by a clock.

This construction was described sixty years ago by Wittgenstein:

> We cannot compare any process with the 'passage of time' — there is no such thing — but only with another process (say, with the movement of a chronometer).
>
> Hence the description of the temporal sequence of events is only possible if we support ourselves on another process. (Wittgenstein, 1933, par.6.3611)

Conclusion

All of this, I believe, is implicit in Piaget's account of the construction of reality. Starting from another kind of analysis, I used a language that is different from his, but came to the very same conclusion. He spoke of a Copernican revolution that prompts the child to construct a world of *external* things, i.e., things beyond the realm of immediate experience (Piaget, 1967b, p.9). It is the world of space and time in which the concept of individual identity assures the continuity of all the perceptual objects we know, but are not attending to at the moment. In short, this construction creates a world of *being*.

The few examples I have given here are by no means representative of all the conceptual analyses that have been carried out; and they are but a small fraction of the multitude that could be analysed. Yet I hope that the sample opens a perspective on a type of investigation that is well worth pursuing because it yields powerful models for the function of both language and thought.

To conclude, I want to emphasize again that conceptual analysis pertains to conceptual structures. This is to say, it pertains to knowledge and not to any reality presumed to be independent of a knower. The concepts of change and state, of space and time, and of a world in which things can perdure and 'exist' while we do not focus attention on them, all these are tools the cognitive subject uses to organize and manage the flow of experience. They cannot reflect the ontological reality of which traditional philosophers dream. Radical constructivism does not speak to this dream; its purpose is to show that a relatively stable 'experiential reality' can be built up without presupposing an independent world-in-itself.

Notes

1   Some of the ideas discussed in this chapter were presented at the twelfth Advanced Course, Archives Jean Piaget, Geneva, September 1992 (Glasersfeld, 1993). Courtesy Archives Jean Piaget.
2   Skinner said this as clearly as one could wish: 'The variables of which human behaviour is a function lie in the environment' (1977, p.1).
3   Note that, in order to remain hidden, the squirrel actually has to move around the tree in the same direction as the man.
4   Though we were taught this method by Ceccato in 1960, he published a description of it only much later (1980, pp.179–82).

Chapter 5

Reflection and Abstraction[1]

Shortly before the turn of the century, John Dewey wrote: 'As adults we are constantly deceiving ourselves in regard to the nature and genesis of our mental experiences' (McLellan and Dewey, 1908, p.27). Much of his work aimed at exposing the deceptions. But the trend in psychology moved in another direction. What came was the behaviourist era. One of its remarkable features is that so many leaders and followers of that creed could claim to be empiricists, cite John Locke as their forefather, and get away with it. Had they read no further than the first chapter of Book II of his major work, *An Essay Concerning Human Understanding*,[2] they would have found some startling things. Right at the beginning there is a caution that might have made them a little more circumspect:

> The understanding, like the eye, whilst it makes us see and perceive all other things, takes no notice of itself; and it requires art and pains to set it at a distance and make it its own object. (John Locke, 1690, Introduction, par.1)

Then, at the beginning of Book II, Locke makes it very clear that he does not intend to do without the 'mind' and its power of 'reflection'. Paragraph 2 has the heading: 'All Ideas come from Sensation or Reflection', and paragraph 4 is entitled 'The operations of our Minds'. It is there that Locke explains what he means by these terms:

> By reflection then, in the following part of this discourse, I would be understood to mean, that notice which the mind takes of its operations, and the manner of them, by reason whereof there come to be ideas of these operations in the understanding. (Locke, 1690, Book II, Chapter I, par.4)

In our century, it was Jean Piaget who vigorously defended and expanded the notion of reflection. He lost no opportunity to distance himself from empiricists who denied the mind and its operations and wanted to reduce all knowing to a passive reception of objective sense data. Yet, he should not have found it difficult to agree with Locke's division of ideas because it is not too different from his own division between figurative and operative knowledge. Both men, I have no doubt, would have agreed with Dewey about the risk of deceiving oneself by taking mental experiences as given. It is therefore with caution that I shall proceed to discuss, in the pages that follow, first my own view of reflection, abstraction, re-presentation, and the use of symbols, and then a tentative interpretation of Piaget's view of reflection.

Reflection

If someone, having just eaten an apple, takes a bite out of a second one, and is asked which of the two tasted sweeter, we should not be surprised that the person could give an answer. Indeed, we would take it for granted that under these circumstances any normal person could make a relevant judgment. We cannot observe how such a judgment is made. But we can hypothesize some of the steps that seem necessary to make it. The sensations that accompanied the eating of the first apple would have to be remembered, at least until the question is heard.[3] Then they would have to be re-presented and compared (in regard to whatever the person called 'sweetness') with the sensations accompanying the later bite from the second apple. This re-presenting and comparing is a way of operating that is different from the processes of sensation that supplied the material for the comparison. Reflecting upon experiences is clearly not the same as *having* an experience.

In 1795, a hundred years after Locke, Wilhelm von Humboldt jotted down a few aphorisms which, posthumously, his editors put under the heading 'About Thinking and Speaking'. The first three aphorisms deal with reflection:

1 The essence of thinking consists in reflecting, i.e., in distinguishing what thinks from what is being thought.
2 In order to reflect, the mind must stand still for a moment in its progressive activity, must grasp as a unit what was just presented, and thus posit it as object against itself.
3 The mind then compares the units, of which several can be created in that way, and separates and connects them according to its needs. (Humboldt, 1907, p.581)[4]

I know of no better description of the mysterious capability that allows us to step out of the stream of direct experience, to re-present a chunk of it, and to look at it as though it were direct experience, while remaining aware of the fact that it is not. I call it mysterious, because, although we can all do it as easily as flipping a switch, we have not even the beginnings of a model (least of all an information processing model) that would suggest how it might be achieved. 'To grasp as a unit what was just presented' is to cut it out of the continuous experiential flow. In the literal sense of the term, this is a kind of abstraction — namely the simplest kind. Focused attention picks a chunk of experience, isolates it from what came before and from what follows, and treats it as a closed entity. For the mind, then, 'to posit it as object against itself', is to re-present it. In the next two sections, I want to deal with abstraction and re-presentation one after the other.

Abstraction

As von Humboldt stated in his third aphorism, chunks of experience, once isolated, can be compared, separated, and connected. This makes possible further steps of abstraction, among them are the kind that Piaget and many others have called 'generalizing abstraction'. Because generalization seems crucial in all forms of naming and categorization, it has been discussed for a long time. To clarify the core of the notion, I once more return to Locke, because he produced a very simple and widely accepted description of the process:

> This is called Abstraction, whereby ideas taken from particular beings become general representations of all the same kind; and their names general names, applicable to whatever exists conformable to such abstract ideas. (Locke, 1690; Book II, Ch.X, par.9)

Locke's use of the words 'being' and 'exist' in this context caused Berkeley, who had a very different view of 'existence', to voice a sarcastic objection against his predecessor.

> Whether others have this wonderful faculty of abstracting their ideas, they best can tell; for myself, I find indeed I have a faculty of imagining, or representing to myself, the ideas of those particular things I have perceived, and of variously compounding them. I can imagine a man with two heads, or the upper parts of a man joined to the body of a horse, I can consider the hand, the eye, the nose, each by itself abstracted or separated from the rest of the body. But then whatever hand or eye I imagine, it must have some particular shape and colour. (Berkeley, 1710, Introduction, par.10)

This passage is interesting for two reasons. Berkeley claims, much as later did von Humboldt, that we are able to represent to ourselves particular experiential items and that we are also able to segment them and to recombine the parts at will. Then however, he goes on to claim that whatever we represent to ourselves mush have the 'character' of a particular, and he concludes that we cannot have general ideas.

Both these claims concern re-presentation and are, I believe, perfectly valid. Whatever you re-present to yourself, be it fish, fowl, or flower, it will be a particular fish, a particular fowl, and a particular flower. It will be of its individual shape, hue, and size, not a 'wild card' that would fit any member of the respective class. What follows from this, is that we are unable to visualize ideas of generalized things, but it does not preclude that we construct general ideas in order to classify particular things. Berkeley, it seems, somehow trapped himself into his position about abstraction although a way out was well within his reach. At the beginning of his treatise, he says among other things:

> Thus, for example, a certain colour, taste, smell, figure and consistence having been observed to go together, are accounted one distinct thing, signified by the name apple; other collections of ideas constitute a stone, a tree, a book, and the like sensible things. (Berkeley, 1710; par. 1)

As he used the indefinite article in 'a stone, a tree, etc.', he was clearly aware of the fact that we apply the name 'apple' not only to one unique thing, but to countless others that fit the description in terms of colour, taste, smell, shape, and consistence. But to him this generality arose from the word and not by *abstracting* the idea from particular instantiations (1710, Introduction, par. 12). Had he analysed it the way he analysed other conceptual operations, he might have changed his view about abstraction. I hope to make this clear with the help of an example.

Generalization

A child growing up in a region where apples are red would necessarily and quite correctly associate the idea of redness with the name 'apple'. A distant relative arriving from another part of the country, bringing a basket of yellow apples, would cause a major perturbation for the child, who might want to insist that yellow things should not be called 'apples'. However, the social pressure of the family's usage of the word would soon force the child to accept the fact that the things people call 'apple' come in different colours. The child might even be told that apples can also be green. This would enable the child to *recognize* as an apple a green thing that satisfied the other relevant conditions the first time it is brought to the house.

Berkeley was, indeed, quite right when he maintained that every time we imagine an apple, it has to have a specific colour. However, this did not justify the claim that we could not abstract, from apple experiences, a general idea that then allows us to recognize as apples items that differ in some respects, but are nevertheless included in that class. The point he missed was that such general ideas are not 'figurative' but 'operative'. That is to say, they are not images like picture postcards but operational recipes that can produce them.

Hence I suggest that we are quite able to abstract general ideas from experience. We do this by substituting a kind of place-holder or variable for some of the properties in the compound sensory structures we actively build up to form particular things from the flow of experience. I see no reason why the resulting operational structure that has the function of a generative programme, should not be called a concept. Such a structure is more specific with regard to some properties and less specific with regard to others; and it is precisely because of this relative indeterminacy that it enables us to recognize items that we have never seen before, as exemplars of a familiar kind.

In short, in order to recognize several particular experiential items as belonging to the same kind, *in spite* of differences they may manifest, we must have a concept that is flexible enough to allow for a certain variability. That is, instead of specific particulars it must contain variables for certain properties. Yet, in order to 'imagine', for instance, an apple, we have to decide what colour it is to be, because we cannot possibly visualize it red and green and yellow at one and the same time. Berkeley, therefore, was right when he observed that whenever we re-present a concept to ourselves, we find that it is a particular thing and not a general idea. What he did not realize was that the abstracted operational pattern necessary to recognize things of a kind, does not automatically turn into an image that can be re-presented.

We shall return to this difference between operational patterns and re-presentations in the context of symbols and language. First, however, I want to clarify the notion of re-presentation.

The Notion of Re-presentation

There are two points I want to make about the term as it is used in the traditional literature, especially in combinations such as 'representational knowledge'. The first point is logical, the second semantic.

At the time of the pre-Socratics, when our epistemological tradition began, it was already clear to some thinkers that a conception of knowledge that required correspondence to a real world was illusory, because there was no way of checking any such correspondence. These thinkers saw with admirable clarity that, in order to judge the goodness of a representation that is supposed to depict something else, one would have to compare it to what it is supposed to represent. In the case of 'knowledge' this would be impossible, because we have no access to the 'real' world except through experience and yet another act of knowing — and this, by definition, would simply yield another representation. Thus there is no difficulty in generating and comparing representations. It is logically impossible, however, to compare a representation with something it is supposed to depict, if that something is supposed to exist in a real world that lies beyond our experiential interface.

William James (1912) neatly formulated the genesis of the paradoxical situation:

> Throughout the history of philosophy the subject and its object have been treated as absolutely discontinuous entities; and thereupon the presence of the latter to the former, or the 'apprehension' by the former of the latter, has assumed a paradoxical character which all sorts of theories had to be invented to overcome. Representative theories put a mental 'representation,' 'image,' or 'content' into the gap, as a sort of intermediary. Common-sense theories left the gap untouched, declaring our mind able to clear it by a self-transcending leap. Transcendentalist theories left it impossible to traverse by finite knowers, and brought an absolute in to perform the saltatory act. (James, 1912, p.27)

My second, semantic, point pertains to the word 'representation' and how it has come to be used in English. Like many other words, it has different meanings. Speakers of the language usually handle ambiguity quite well; but in the case of this particular word there is a peculiar difficulty: one of its ambiguities seems to have sprung, not from the word's original use in English, but from an unfortunate use introduced, it seems, by translators of German philosophy. It may have started earlier, but it became common usage in philosophy with the translation of Kant's *Critique of Pure Reason*. The two German words *Vorstellung* and *Darstellung* were rendered by one and the same English word 'representation'. To speakers of English this implies a reproduction, copy, or other structure that is in some way isomorphic with an original. This condition fits the second German word quite well, but it does not fit the first. *Vorstellung*, which is the word Kant uses throughout his work, should have been translated as 'presentation', because it designates, among other things, the 'performance' of a magician, and one would use it to ask a theatre: How many 'shows' are there on Saturday?.

The conflation of the two concepts is obviously disastrous in epistemological contexts. Although both the German words are used to refer to conceptual structures, they specify incompatible characteristics. The element of autonomous construction is an essential part of the meaning of *Vorstellung*. If it is lost, one of the most important features of Kant's (and Piaget's) theory becomes incomprehensible.

Re-presenting Past Experiences

There is no doubt that the human mind can re-present things to itself that have been, or are not yet, actual experiences. Though I have not the vaguest idea *how* I do it, I can at this moment re-present to myself the way up a mountain I climbed on a winter's day, forty years ago in the Swiss Alps. I can hear that peculiar swishing, crunching sound at each step, as I push a ski forward into untouched snow and then put my weight on it. I can see the track I am making, in front of me as a project, behind me as a product, as it follows the contour of slopes and gullies, and I can feel that constant effort to keep the track at a steady gradient; and I can smell, with every breath, that incomparable combination of dry, cold air and brilliant sunlight.

It is clear that in this context, 'to hear', 'to see', 'to feel', and 'to smell' do not refer to quite the same activities as in a context of immediate perception. When I perceive, I would say I am registering signals that seem to come from my eyes, ears, and nose. When I re-present something to myself, it seems to come from another source, a source that feels as though it were wholly inside. Perhaps this difference springs largely from the experiential fact that, when I perceive, my percepts can be modified by my physical motion. The past I re-present to myself, in contrast, is not influenced by the way I move at present.

As I said, I do not know how re-presentation works. In fact, no one, today, knows how it works. We have not even the beginnings of a plausible functional model of human memory, let alone a model of human consciousness. Yet, something we want to call memory as well as something we want to call consciousness are involved in the kind of re-play of past experiences that I was describing. The point I want to make is this: If I re-present to myself something that was a familiar experience forty years ago, it is, indeed, very much like re-playing or reconstructing something that was experienced at another time. More important still, it is under all circumstances a re-play of *my own* experiences, not a piece of some independent, objective world.

That is the reason why I insist on the hyphen. I want to stress the 're-' because it brings out the repetition — repetition of something that was present in a subject's experiential world at some other time.[5]

In general then, re-presentation spelled with a hyphen, is intended as a mental act that brings a prior experience to an individual's consciousness. More specifically, it is the recollection of the figurative material that constituted the experience. No such recollection would be possible if the original generation of the experience had not left some trace or mark to guide its reconstruction.

Recognition

In requiring memory, recognition is similar to re-presentation. Both often work hand in hand, as for instance, when one recognizes a Volkswagen though one can see only part of its back but is nevertheless able to visualize the whole characteristic shape. However, the ability to recognize a thing from a partial presentation in one's perceptual field, does not necessarily bring with it the ability to re-present the thing spontaneously. We have all had occasion to notice this. Our experiential world contains many items which, although we recognize them when we see them, are not available to us when we want to visualize them. There are, for instance, people whom we would recognize as acquaintances when meeting them face to face, but were we asked to describe them when they are not in our visual field, we would be unable to recall an adequate image of their appearance.

The fact that recognition developmentally precedes the ability to re-present an experiential item spontaneously, has been observed in many areas. It is probably best known and documented as the difference between what linguists call 'passive' and 'active' vocabulary. The difference is conspicuous in second-language learners but it is noticeable also in anyone's first language: a good many words one knows when one hears or reads them, are not available when one is speaking or writing.

This developmental lag suggests that to re-present a perceptual item to oneself in its absence, requires something more than the conceptual structure that serves to recognize it. Piaget has always maintained that all forms of imaging and re-presenting are, in fact, acts of internalized imitation (Piaget, 1945).[6]

However, there is a difference between imitating something that one has just constructed out of material that is still present (I would call this 'copying') and imitating something from memory — just as it is more difficult to draw something remembered than to draw it from life. A computer programme and a map are useful metaphors to bring out the added difficulty.

The Need of an Agent

A programme embodies the fixed itinerary of a given activity and therefore can guide and govern its re-enactment. But a programme can only specify the material on which to act. It does not supply the material; nor does it supply the acting agent and the performance of the action.

An analogous limitation, I suggest, may account for the fact that to recognize an experiential item requires less effort than to re-present it spontaneously. This would be so, because in re-presentation not only a programme of composition is needed, but also the specific sensory components, which must be expressly generated. In recognition, the perceiver merely has to isolate the particular elements in the sensory manifold. As Berkeley observed, sensory elements are 'not creatures of the will' (1710 par.29). Because there are always vastly more sensory elements than the perceiving agent can attend to and use, recognition requires the attentional selecting, grouping, and coordinating of sensory material that fits the composition programme of the item to be recognized.[7] In re-presentation, on the other hand, some substitute for the sensory raw material must be generated. (As the example of the Volkswagen indicates, the re-generation of sensory material is much easier when parts of it are supplied by perception, a fact that was well known to the proponents of Gestalt psychology.)

In some respects, a programme is like a map. If someone draws a simple map to show you how to get to his house, he essentially indicates a potential path to the unknown location, starting from a place you are presumed to know. The drawing of the path is a graphic representation of turns and where they have to be made to accomplish the desired itinerary. It does not and could not indicate the active agent that has to supply the movement, nor does it indicate what it means to turn right or left. Any user of the map, must supply the motion and the changes of direction with the focus of visual attention while reading the map. Only if one manages to abstract this sequence of motions from the reading activity, can one transform it into physical movement through the mapped region. (Note that this abstracting and transforming into physical movement is by no means an easy task for people who are unaccustomed to map reading.)

The programme, however, differs from a map in that it explicitly provides instructions about actions and implicitly indicates changes of location through the conventional sequence in which the instructions must be read. As in the map, it is the user's focus of attention which, being an integral part of the reading or implementing, supplies the progressive motion. But, unlike a map, a programme may contain embedded subroutines. Yet, no matter how detailed these subroutines might be, they can contain only instructions to act, not the actions themselves. In other words, irrespective of how minutely a programme's instructions have decomposed an activity, they remain static until some agent implements them and adds the dynamics.

In carrying out a programme in an experiential situation, just as in following a map through an actual landscape, the sensory material in the agent's perceptual field can supply cues as to the action required at a given point of the procedure. In the re-presentational mode, however, attention cannot focus on actual perceptual material and pick from it cues about what to do next, because the sensory material itself has to be generated. A re-presentation — at least when it is a spontaneous one — is wholly self-generated (which is one reason why it is usually easier to find one's way through a landscape than to draw a reliable map of it when one is not there.)

The increase of difficulty and the concomitant increase of effort involved in the production of conceptual structures when the required sensory material is not available in the present perceptual field, shows itself in all forms of re-presentation and especially in the re-enactment of abstracted programmes of action. Any re-presentation, be it of an experiential object or of a programme of actions or operations, requires *some* sensory material for its execution. This basic condition, I believe, is what confirmed Berkeley in his argument against the 'existence' of abstracted general ideas; for it is indeed the case that every time we re-present to ourselves such a general idea, it turns into a particular one because its implementation requires the kind of material from which it was abstracted.

This last condition could be reformulated by saying that there has to be some isomorphism between the present construct and what it is intended to re-present. Clearly, this isomorphism does not concern a thing-in-itself but those aspects of a past experience one wants (or happens) to focus on. In this context I always remember a shrewd observation Silvio Ceccato made. When we dream, he said, we operate in the opposite direction of perception: we begin with concepts of objects and visualize no more of their perceptual features than are required by the story of the dream.[8]

Meaning as Re-presentation

More importantly, this selective isomorphism is the basis of graphic and schematic representations (without hyphen!). They tend to supply such perceptual material as is required to bring forth in the perceiver the particular ways of operating that the maker of the graphic or schematic is aiming at. In this sense they are didactic (Kaput, 1991), because they can help to focus the naive perceiver's attention on the execution of the particular operations that are deemed desirable. Hence, as I have suggested elsewhere (see Glasersfeld, 1987), graphic representations may be iconic (i.e., picture-like) or symbolic, but neither kind should be confused with the mental re-presentations I am discussing here.

Re-presentations can be activated by many things. Any element in the present stream of experience may bring forth the re-presentation of a past situation, state, activity, or other construct. This experiential fact was called 'association' by the early empiricist philosophers and Freud took it as the basis of his analyses of neuroses. The ability to associate is systematically exploited by language. To *know* a word is to have a meaning associated with it. The meaning may be figurative, (abstracted from sensorimotor experience), operative (indicating a conceptual relation or other mental operations), or a complex conceptual structure involving both figurative and operative elements.[9]

Figurative meanings are those that can be visualized immediately because they call up a re-presentation of the kind of sensorimotor experience from which they were abstracted. These re-presentations, moreover, are often incomplete. For example, we can all visualize ourselves crawling on all fours, but Piaget demonstrated that, in order to be sure of the sequence in which we move arms and legs, many adults must actually carry out the activity (1974a, p.15).

Operative meanings cannot be re-presented as such but only as 'implemented' in a sensorimotor situation. A golfer, for example, cannot recall a 'swing' without re-presenting to himself the feel of the club in his hands, the kinaesthetic signals from the muscles in his arms, his stance on the ground, and other sensory material.

The language user therefore has to assume that whatever re-presentation he or she has associated with a word is somehow similar to the re-presentations the word brings forth in other users of the language. The assumption of some such parallelism is the foundation of what is commonly called 'communication'. The claim that these re-presentations are shared by all speakers of a language is naive and unwarranted, because all that can ever be shown is that the individuals' re-presentations are compatible in the given context.

The Power of Symbols

A detailed model of the semantic dimension of words will be presented in Chapter 7. Here, in order to discuss the role of recognition and re-presentation in the use of words, two simple but indispensable conditions have to be explained. The first is that the phonemes that compose the word in speech, or the graphic marks that constitute it in writing, must be recognized as that particular item of one's vocabulary. As I mentioned earlier, this recognizing ability, is developmentally prior to the ability to re-present and produce the word spontaneously.

If the word merely causes a response in the form of an action, for instance when it is used as command, I call it a 'signal'. If a second condition is satisfied, I will say that the word is used as a 'symbol'.[10] In my terminology, a word will be considered a symbol, only when it brings forth in the user an abstracted re-presentation. The word/symbol, therefore, must be associated with a conceptual structure that was abstracted from experience and, at least to some extent, generalized.

Once a word has become operative as a symbol and calls forth the associated meaning as re-presented chunks of experience that have been isolated (abstracted), its power can be further expanded in an important way. As individual users of the word become more proficient, they no longer need to actually produce the associated conceptual structures as a completely implemented re-presentation. They may simply register the occurrence of the word as a kind of 'pointer' to be followed if needed at a later moment. I see this as analogous to the capability of recognizing objects on the basis of a partial perceptual construction. In the context of symbolic activities, this capability is both subtle and important. An example may help to clarify what I am trying to say.

If, in someone's account of a European journey, you read or hear the name 'Paris', you may register it as a pointer to a variety of experiential referents with which you happen to have associated it — e.g., a particular point on the map of Europe, your first glimpse of the Eiffel Tower, the Mona Lisa in the Louvre — but if the account of the journey immediately moves to London, you would be unlikely to implement fully any one of the relevant experiences as an actual re-presentation. At a subsequent moment, however, if the context or the conversation required it, you could return to the mention of Paris and develop one of the associated re-presentations.

I have chosen to call this function of symbols 'pointing', because it seemed best to suggest that words/symbols acquire the power to open or activate pathways to specific re-presentations without, however, obliging the proficient symbol user to produce the re-presentations there and then.

This function, incidentally, constitutes one of the central elements of our theory of children's acquisition of the concept of number (Steffe, Glasersfeld, Richards and Cobb, 1983). In this theory, the first manifestation of an abstract number concept is a demonstration that the subject knows, without carrying out a count, that a number word implies or points to the sequential one-to-one coordination of all the terms of the standard number-word sequence from 'one' up to the given word (and that all of them can be coordinated with some countable items). Indeed, we believe that this is the reason why, as adults, we may assert that we know what, say, the numeral (symbol) '381,517' means, in spite of the fact that we are unlikely to be able to re-present to ourselves a collection of that many discrete experiential items. We know what the number word means, because it points to the last element in a familiar counting procedure (or other mathematical method).

In mathematics this form of symbolic implication is so common that it usually goes unnoticed. For instance, when you read that the side of a pentagon is equal to half the radius of the circumscribed circle multiplied by $\sqrt{10 - 2\sqrt{5}}$, you do not have to draw the square roots to understand the statement — provided you know the operations the symbols point to. The potential ability is sufficient, the actual operations do not have to be carried out. Because it is so often taken for granted that mathematical expressions can be understood without carrying out the operations they symbolize, formalist mathematicians are sometimes carried away and declare that the manipulation of symbols constitutes mathematics. Without the mental operations they indicate, however, symbols are reduced to meaningless marks (see Hersh, 1979, p.19).

Piaget's Theory of Abstraction

Few, if any, thinkers in this century have used the notion of abstraction as often and insistently as Piaget did. Indeed, in his view 'All new knowledge presupposes an abstraction, . . .' (Piaget, 1974c, p.81). But not all abstractions are the same. Piaget distinguished two main kinds, 'empirical' and 'reflective', and then subdivided the second. He has frequently explained the primary difference in seemingly simple terms, for example:

> Empirical abstractions concern observables and reflective abstractions concern coordinations. (Piaget *et al.*, 1977a, Vol. 2; p.319)

> One can thus distinguish two kinds of abstraction according to their exogenous or endogenous sources; . . . (Piaget, 1974c, p.81)

Anyone who has entered into the spirit of genetic epistemology will realize that the simplicity of these statements is deceptive. The expressions 'observables' and 'exogenous' are liable to be interpreted in a realist sense, as aspects or elements of an external reality. Given Piaget's theory of knowledge, however, this is not how they were intended. In fact, the quoted passages are followed by quite appropriate warnings. After the first, Piaget explains that no characteristic is in itself observable. Even in physics, he says, the measured magnitudes (mass, force, acceleration, etc.) are themselves constructed and are therefore results of inferences deriving from preceding abstractions (op. cit.). In the case of the second quotation, he adds a little later: 'there can be no exogenous knowledge except that which is grasped as content, by way of forms which are endogenous in origin.' (Piaget, 1974c, p.83). This is not an immediately transparent formulation. As so often in Piaget's writings, one has to look elsewhere in his work for enlightenment.

Form and Content

The distinction between form and content has a history as long as western philosophy and the terms have been used in many different ways. Piaget's use of the distinction is complicated by the fact that he links it with his use of 'observables' (content) and 'coordinations' (forms).

> The functions of form and content are relative, since every form becomes content for another that comprises it. (Piaget *et al.*, 1977a, Vol 2; p.319).

This will make sense, only if one recalls that, for Piaget, percepts, observables, and any knowledge of objects, are all the result of a subject's action and not externally caused effects registered by a passive receiver. In his theory, to perceive, to remember, to re-present, and to coordinate are all *dynamic*, in the sense that they are activities carried out by a subject that operates on internally available material and produces certain results.

A term such as 'exogenous', therefore, must not be interpreted as referring to what is supposed to be physically outside relative to a physical organism, but rather as referring to something that is external relative to the process in which it is about to become involved.

Observation and re-presentation have two things in common:

- They operate on items which, relative to the process at hand, are considered given. The present process takes them as elements and coordinates them as 'content' into a new 'form' or 'structure'.
- The resulting new products can be taken as initial 'givens' by a future process of structuring, relative to which they then become 'content'. Thus, once a process is achieved, its results may be considered 'observables' or 'exogenous' relative to a subsequent process of coordination or a higher level of analysis.

As Piaget saw, this might seem to lead to an infinite regress (ibid., p.306), but he put forth at least two arguments to counter this notion. One of them emerges from his conception of scientific analysis. Very early in his career, he saw this analysis as a cyclical programme in which certain elements abstracted by one branch of science become the givens for coordination and abstraction in another. In an early paper (Piaget, 1929) and almost forty years later in his 'classification of the sciences' (Piaget, 1967a, 1967c), he formulated this mutual interdependence of the scientific disciplines as a closed cycle of changes: biology→psychology→mathematics→physics, and looping back to biology. From his perspective, there is no linear progression without end, but simply development of method and concepts in one discipline leading to novel conceptualization and coordination in another. The recent impact of the physics of molecules and particles on the conceptual framework of biology would seem a good example (but it is no doubt too much to expect that in the near future research grants will be offered for the study of how the concepts of modern physics developed from psychological gambits with the help of mathematical concepts).

The second reason against an infinite regress of abstractions is grounded in the developmental basis of genetic epistemology and is directly relevant here. The child's cognitive career has an unquestionable beginning, a first stage during which the infant assimilates, or tries to assimilate, all experience to such fixed action patterns (reflexes) as it has at the start (Piaget, 1975, p.180). Except for their initial fixedness, these action patterns function like the schemes which the child begins to coordinate a little later on the basis of expanding experience (see Chapter 3).

Early in the sensorimotor period in infancy (i.e., the child's first two years), assimilation and accommodation are assumed to take place without awareness and conscious reflection. The fact that 3 or 4-month-old infants assimilate items (which, to an observer, are not all the same) as triggers of a particular scheme, is sometimes described as the ability to generalize (animal psychologists, working with rats or monkeys, call it 'stimulus generalization'). A little later in the sensorimotor stage, reflection begins to operate and with it the discrimination of experiential items that do function in a given scheme, from others that do not. Thus a mechanism is initiated that furnishes the source of empirical abstractions which, in turn, lead to the child's ability to re-present experienced items to him- or herself when they are not actually present. This inevitably raises the question when and how the acting subject's awareness is involved.

It is an urgent question because the word 'reflection', ever since Locke introduced it into the human sciences, tends to imply a conscious mind that does the reflecting. A second reason is that in many places where Piaget draws the distinction between the 'figurative' and the 'operative', it tends to rein-force the notion that the operative (described by both Locke and Piaget as 'the ideas the mind gleans by reflecting on its own operations') requires consciousness. Consequently, it would be desirable to unravel when, in Piaget's theory of cognitive development, the capability of conscious reflection arises.

Piaget himself, as I have said elsewhere (Glasersfeld, 1982), rarely makes explicit whether, in a given passage, he is interpreting what he is gathering from his observations (observer's point of view), or whether he is conjecturing an autonomous view from the observed subject's perspective (see Vuyk, 1981, Vol.2). This difference seems crucial in building a model of mental operations and, therefore, to an understanding of his theory of abstraction and, especially, reflective abstraction. I shall return to this question of consciousness after the next section. First I shall try to lay out the different kinds of abstraction Piaget has distinguished.

Four Kinds of Abstraction

The process Locke characterized by saying, 'whereby ideas taken from particular beings become general representations of all the same kind', (1690, Book II, Ch.X, par.9) falls under Piaget's term 'empirical abstraction'. To isolate certain sensory properties of an experience and to maintain them as repeatable combinations, i.e., isolating what is needed to recognize further instantiations of, say, apples, undoubtedly constitutes an empirical abstraction. But, as I suggested earlier, to have composed a concept that can serve to recognize (assimilate) items as suitable triggers of a particular scheme, does not automatically bring with it the ability to visualize such items spontaneously as re-presentations.

Piaget makes an analogous point — incidentally, one of the few places where he mentions an empiricist connection:

> But it is one thing to extract a character, x, from a set of objects and to classify them together on this basis alone, a process which we shall refer to as 'simple' abstraction and generalisation (and which is invoked by classical empiricism), and quite another to recognise x in an object and to make use of it as an element of a different (non-perceptual) structure, a procedure which we shall refer to as 'constructive' abstraction and generalisation. (Piaget, 1969, p.317)

The capability of spontaneous re-presentation develops in parallel with the acquisition of language and may lead to an initial, albeit limited form of awareness. Children at the age of 3 or 4 years, are not incapable of producing some pertinent answer when they are asked what a familiar object is like or not like, even when the object is not in sight at the moment. This suggests that they are able not only to call forth an empirically abstracted re-presentation but also to review it quite deliberately.

The notion of empirical abstraction covers a wider range of experience for Piaget than is envisioned in the passage I quoted from Locke. What Locke called 'particular beings' were for him *ideas* supplied by the five senses. Because, in Piaget's view, visual and tactual perception involve motion, it is not surprising that the internal sensations caused by the agent's own motion (kinesthesis) belong to the 'figurative' and are therefore, for him, raw material for empirical abstractions in the form of motor patterns.[11]

That such abstracted motor patterns reach the level where they can be represented, you can check for yourself. Anyone who has some proficiency in activities such as running down stairs, 'serving' in tennis, swinging for a drive in golf, or skiing down a slope, has no difficulty in re-presenting the involved movements without stirring a muscle. An interesting aspect in such 'dry reruns' of abstracted kinaesthetic experiences is that they don't require specific staircases, balls, or slopes. I mention this because it seems to me to be a clear demonstration of deliberate and therefore conscious re-presentation of something that needed no consciousness for its abstraction from actual experience. This difference is relevant to the subdivisions Piaget introduced in the area of reflective abstraction.

From empirical abstractions, which have sensorimotor experience for raw material, Piaget, as I said earlier, distinguished three types of reflective abstraction. Unfortunately, the French labels Piaget chose for them are inevitably confused by literal translation into English.

The first reflective type derives from a process Piaget calls *réfléchissement*, a word that is used in optics when something is being reflected, as for instance the sun's rays on a shiny surface. In his theory of cognition, the term is used to indicate that an activity or mental operation (not a static combination of sensory elements) developed on one level is abstracted from that level of operating and applied to a higher one, where it is then considered to be a *réfléchissement*. Moessinger and Poulin-Dubois, 1981, have translated this as 'projection', which captures something of the original sense. But Piaget stresses that a second characteristic is required:

> Reflective abstraction always involves two inseparable features: a *réfléchissement* in the sense of the projection of something borrowed from a preceding level onto a higher one, and a *réflexion* in the sense of a (more or less conscious) cognitive reconstruction or reorganisation of what has been transferred. (Piaget, 1975, p.41)

At the beginning of the first of his two volumes on reflective abstraction (Piaget *et al.*, 1977a), the two features are again mentioned:

> Reflective abstraction, with its two components of *réfléchissement* and *réflexion*, can be observed at all stages: from the sensorimotor levels on, the infant is able, in order to solve a new problem, to borrow certain co-ordinations from already constructed structures and to reorganise them in function of new givens. We do not know, in these cases whether the subject becomes aware of any part of this. (Piaget *et al.*, 1977a, Vol 1; p.6)

In the same passage he immediately goes on to describe the second type of reflective abstraction:

In contrast, at the later stages, when reflection is the work of thought, one must also distinguish thought as a process of construction and thought as a process of retroactive thematisation. The latter becomes a reflecting on reflection; and in this case we shall speak of *abstraction réfléchie* (reflected abstraction) or *pensée réflexive* (reflective thought). (op. cit.)

Since the present participle of the verb *réfléchir*, from which both the nouns *réfléchissement* and *réflexion* are formed, is *réfléchissante*, Piaget used *abstraction réfléchissante* as a generic term for both types. It is therefore not surprising that in most English translations the distinction was lost when the expression 'reflective abstraction' was introduced as the standard term.

The situation is further compounded by the fact that Piaget distinguished a third type of reflective abstraction which he called 'pseudo-empirical'. When children are able to re-present certain things to themselves, but are not yet fully on the level of concrete operations,

it happens that the subjects, by leaning constantly on their perceivable results, can carry out certain constructions which, later on, become purely deductive (e.g. using an abacus or the like for the first numerical operations). In this case we shall speak of 'pseudo-empirical abstraction' because, in spite of the fact that these results are read off material objects as though they were empirical abstractions, the perceived properties are actually introduced into these objects by the subject's activities. (Piaget, ibid.)

To recapitulate, Piaget distinguishes four kinds of abstraction. One is called 'empirical' because it abstracts sensorimotor properties from experiential situations. The first of the three reflective abstractions projects and reorganizes, on another conceptual level, a coordination or pattern of the subject's own activities or operations. The next is similar in that it also involves patterns of activities or operations, but it includes the subject's awareness of what has been abstracted and is therefore called 'reflected abstraction'. The last is called 'pseudo-empirical' because, like empirical abstractions, it can take place only if suitable sensorimotor material is available.

The Question of Awareness

One of the two main results of the research carried out by Piaget and his collaborators on the attainment of awareness, he summarized as follows in *La prise de conscience*:[12]

> ... action by itself constitutes an autonomous knowledge of consid-
> erable power, for while it is only 'know-how' and not knowledge
> that is conscious of itself in the sense of conceptualised understand-
> ing, it nevertheless constitutes the source of the latter, because the
> attainment of consciousness nearly always lags quite noticeably be-
> hind this initial knowledge which is remarkably efficacious even though
> it does not know itself. (Piaget, 1974a, p.275)

The fact that conscious, conceptualized knowledge of a given situation developmentally lags behind the knowledge of how to act in the situation, is commonplace on the sensorimotor level. In my view, as I mentioned earlier, this is analogous to the temporal lag of the ability to re-present a given item relative to the ability to recognize it. But the ability spontaneously to re-present to oneself a sensorimotor image of, say, an apple, still falls short of what Piaget in the above passage called 'conceptualised understanding'. This would involve awareness of the characteristics inherent in the concept of apple (or whatever one is re-presenting to oneself) and this kind of awareness con-stitutes a higher level of mental functioning than mere use of a re-presentation.

This further step requires a good deal more of what Locke called the mind's 'art and pains to set (something) at a distance and make it its own object'. A familiar motor pattern is once more a good example: we may be well able to re-present to ourselves a tennis stroke or a golf swing, but few, if any, would claim to have a conceptualized understanding of the sequence of elementary motor acts that are involved in such an abstraction of a deli-cately coordinated activity. Yet it is clear that, insofar as such understanding is possible, it can be built up only as a 'retroactive thematization', that is, after the whole kinaesthetic pattern has been empirically abstracted from the experience of practising the activity.

In Piaget's theory, the situation is similar in the first type of reflective abstraction: he maintains that it, too, may or may not involve the subject's awareness.

> Throughout history, thinkers have used thought structures without
> having grasped them consciously. A classic example: Aristotle used
> the logic of relations, yet ignored it entirely in the construction of his
> own logic. (Piaget and Garcia, 1983, p.37)

In other words, one can be quite aware of what one is cognitively oper-ating on, without being aware of the operations one is carrying out.

As for the second type, 'reflective thought' or 'reflected abstraction', it is the only one about which Piaget makes an explicit statement concerning awareness:

> Finally, we call the result of a reflective abstraction 'reflected' abstrac-
> tion, once it has become conscious, and we do this *independently of its
> level*. (Piaget *et al.*, 1977a, vol.2; p.303; my emphasis)

When one comes to this statement in Piaget's summary at the end of the second volume on the specific topic of reflective abstraction, it becomes clear that the sequence in which he usually discusses the three types is a little misleading. It is neither a developmental nor a logical sequence. What he appropriately calls 'reflective thought' and lists as the second of three types, describes a cognitive phenomenon that is much more sophisticated than reflective abstractions of type 1 or type 3. Moreover, it is relevant also as a further development of 'empirical' abstraction.

I would suggest that the two meanings of the word 'reflection' be assigned in the following way to Piaget's classification of abstractions: it should be interpreted as projection and adjusted organization on another operational level in the case of reflective abstraction type 1 and pseudo-empirical abstraction; and it should be taken as conscious thought in the case of 'reflected' abstraction, i.e., type 2.

In his two volumes *La prise de conscience* (1974a) and *Réussir et comprendre* (1974b), there is a wealth of observational material from which Piaget and his collaborators infer that consciousness appears hesitantly in small steps, each of which conceptualizes a more or less specific way of operating. Like von Humboldt, Piaget takes the mind's ability to step out of the experiential flow for granted. He then endeavours to map when and under what conditions the subject's awareness of its own operating sets in; and he tries to establish how action evolves in its relation to the conceptualization which characterizes the attainment of consciousness (Piaget, 1974a, p.275ff). In the subsequent volume, he provides an excellent definition of what it is that awareness contributes:

> To succeed is to comprehend in action a given situation to a degree sufficient to attain the proposed goals; to understand is to master in thought the same situations to the point that one can resolve the problems they pose with regard to the why and the how of the links one has established and used in one's actions. (Piaget, 1974b, p.237)

The cumulative result of the minute investigations contained in these two volumes enabled Piaget to come up with an extremely sophisticated description of the mutual interaction between the construction of successful schemes and the construction of abstracted understandings, an interaction that eventually leads to accommodations and to finding solutions to problems in the representational mode, i.e., without having to have run into them on the level of sensorimotor experience.

In this context, one further thing must be added. In the earlier sections, I discussed the fact that re-presentation follows upon recognition and that the 'pointing' function of symbols follows as the result of familiarity with the symbols' power to bring forth re-presentations that are based on empirical abstractions. As the examples I gave of abstracted motor patterns should make clear, symbols can be used, simply to point to such patterns, in which case the re-presentation of action can be curtailed, provided the subject has consciously conceptualized the action and knows how to re-present it.

Operational Awareness

I now want to emphasize that this pointing function of symbols makes possible a way of mental operating that comes to involve conscious conceptualization and, as a result, gives more power to the symbols. Once reflective thought can be applied to the kind of abstraction Piaget ascribed to Aristotle (see passage quoted above), there will be awareness not only of what is being operated on but also of the operations that are being carried out. Piaget suggested this in an earlier context:

> A form is indissociable from its content in perception but can be manipulated independently of its content in the realm of operations, in which even forms devoid of content can be constructed and manipulated . . . logico-mathematical operations allow the construction of arrangements which are independent of content . . . pure forms . . . *simply based on symbols.* (Piaget, 1969, p.288; my emphasis)

In my terms this means, symbols can be associated with operations and, once the operations have become quite familiar, the symbols can be used to point to them without the need to produce an actual re-presentation of carrying them out. If this is accepted as a working hypothesis, we have a model for a mathematical activity that was very well characterized by Juan Caramuel, twenty-five years before Locke published his Essay:[13]

> When I hear or read a phrase such as 'The Saracen army was eight times larger than the Venetian one, yet a quarter of its men fell on the battlefield, a quarter were taken prisoner, and half took to flight', I may admire the noble effort of the Venetians and I can also understand the proportions, without determining a single number. If someone asked me how many Turks there were, how many were killed, how many captured, how many fled, I could not answer unless one of the indeterminate numbers had been determined . . .
>
> Thus the need arose to add to common arithmetic, which deals with the determinate numbers, another to deal with the indeterminate numbers. (Caramuel, 1670/1977, p.37)

In Europe, Caramuel says, this other arithmetic, which deals with abstractions that are 'more abstract than the abstract concept of number', became known as 'algebra'. Given the model of abstraction and reflection I have discussed in these pages, it is not difficult to see what this further abstraction resides in. To produce an actual re-presentation of the operative pattern abstracted from the arithmetical operation of, say, division, specific numbers are needed. This is analogous to the need of specific properties when the re-presentation of, say, an apple, is to be produced. But there is a difference: the properties required to form an apple re-presentation are sensory properties, whereas the numbers needed to re-present an operative pattern in arithmetic are themselves abstractions from mental operations and therefore re-presentable only with the help of some sensory material. Yet, once symbols have been associated with the abstracted operative pattern, these symbols, thanks to their power of functioning as pointers, can be understood, without the actual production of the associated re-presentation — provided the user knows how to produce it when the numerical material is available.

Conclusion

Abstraction, re-presentation, reflection, and conscious conceptualization interact on various levels of mental operating. In the course of these processes, what was produced by one cycle of operations, can be taken as *given* content by the next one, which may then coordinate it to create a new 'form', a new structure. Any such structure can then be consciously conceptualized and associated with a symbol. The structure that then functions for the particular cognizing subject as the symbol's 'meaning', may have gone through several cycles of abstraction and reorganization. This is one reason why the conventional view of language is misleading. In my experience, the notion that word/symbols have fixed meanings that are shared by every user of the language, breaks down in any conversation that attempts an interaction on the level of concepts, that is, a conversation that attempts to go beyond a simple exchange of soothing familiar sounds.

In analyses like those I have tried to lay out in this chapter, one chooses the words that one considers the most adequate to establish the similarities, differences, and relationships one has in mind. But the meanings of whatever words one chooses are one's own, and there is no way of presenting them to a reader for inspection. This, of course, is the very same situation I find myself in, *vis-à-vis* the writings of Piaget. There is no way of discovering what he had in mind — not even by reading him in French. All I — or anyone — could do, is *interpret*, which is to say, construct and reconstruct until a satisfactory degree of coherence is achieved among the conceptual structures one has built up on the basis of the read text.

This situation, I keep reiterating, is no different from the situation we are in, *vis-à-vis* our non-linguistic experience, i.e., the experience of what we like to call 'the world'. What matters there, is that the conceptual structures we abstract turn out to be suitable in the pursuit of our goals; and if they do suit our purposes, they must also be brought into some kind of harmony with one another. This is the same, whether the goals are on the level of sensorimotor experience or of reflective thought. From this perspective, the test of anyone's

All we can do is interpret

account that purports to interpret direct experience or the writings of another, must be whether or not this account brings forth in the reader a network of conceptualizations and reflective thought that he or she finds coherent and useful.

Philosophical Postscript

It may be time for a professional philosopher to re-evaluate the opposition between empiricism and rationalism. The rift has been exaggerated by an often ill-informed tradition in the course of the last 100 years, and the polarization has led to utter mindlessness on the one side and to various kinds of solipsism on the other. Yet, if we return to Locke, from the partially Kantian position of a constructivist such as Piaget, we may be able to reformulate the difference.

> *The Original of all our Knowledge* — In time the mind comes to reflect on its own operations about ideas got by sensation, and thereby stores itself with a new set of ideas, which I call ideas of reflection. These are the impressions that are made on our senses by outward objects that are extrinsical to the mind; and its own operations, proceeding from powers intrinsical and proper to itself, which, when reflected on by itself, become also objects of contemplation — are, as I have said, the original of all knowledge. (Locke, 1690, Book II, Ch.1, par.24)[14]

With one modification, this statement fits well into my interpretation of Piaget's analysis of abstractions. The modification concerns, of course, the 'outward objects that are extrinsical to the mind'. In Piaget's view, exogenous and endogenous do not refer to an inside and an outside relative to the organism, but are intended relative to the mental process that is going on at the moment. The internal construct that is formed by the coordination of sensorimotor elements on one level, becomes external material for the coordination of operations on the next higher level. The only thing Piaget assumes as a given starting-point for this otherwise closed but spiralling process, is the presence of a few fixed action patterns at the beginning of the infant's cognitive development.

Both Locke's and Piaget's model of the cognizing organism acknowledge the senses and the operations of the mind as the two sources of ideas. Locke believed that the sensory source of ideas, the 'impressions' generated by 'outward objects', provided the mind with some sort of picture of an outside world. Piaget saw perception as the result of the subject's actions and mental operations aimed at providing, not a picture of, but an adaptive fit into the structure of that outer world. Consequently, he assigned the functional primacy of the two sources differently. Locke, especially later in his work, tends to emphasize the passive reception of impressions by the senses. Piaget, instead, posits as primary the active mind that organizes sensation to form percepts. The difference, however, takes on an altogether changed character, once we consider that the concept of knowledge is not the same for both thinkers. For Locke it still involved the notion of 'truth' as correspondence to an independent outside world; for Piaget, in contrast, it has the biologist's meaning of functional fit or viability as the indispensable condition of organic survival and cognitive equilibration.

The difference, therefore could be characterized by saying that classical empiricism accepts without question the static notion of being, whereas constructivist rationalism accepts without question the dynamic notion of living.

Notes

1   Revised and expanded from a paper published in *Epistemological foundations of mathematical experience*, edited by L.P. Steffe (1991). (Courtesy Springer Verlag, New York).

2   Locke divided this work into Books, Chapters, and numbered paragraphs.

3   Memory, as Heinz von Foerster (1965) pointed out, cannot be a fixed record (because the capacity of heads, even on the molecular level, is simply not large enough); hence, it must be thought of as dynamic, i.e., as a mechanism that reconstructs rather than stores.

4   A first English translation of von Humboldt's aphorisms was published by Rotenstreich (1974). The slightly different translations given here are mine.

5   Re-presentation may also refer to a new construction (from remembered elements) that has not yet been actually experienced as such, but is projected into the future as a possibility.

6   It is crucial to keep in mind that Piaget emphatically stated that knowledge could not be a copy or picture of an external reality; hence, for him, 'imitation' did not mean producing a replica of an object outside the subject's experiential field, but rather the re-generation of an externalized experience.

7   See William James (1892/1962, p.277): 'One of the most extraordinary facts of our life is that, although we are besieged at every moment by impressions from our whole sensory surface, we notice so very small a part of them.'

8   Ceccato said this in our discussions on the operations that constitute 'meaning' (1947–52).

9   Note that I am using the term 'figurative' in Piaget's sense, not as a synonym of 'transferred sense'.

10   My use of the word 'symbol' follows Susanne Langer (1948) and is not the same as Piaget's for whom symbols had to have an iconic relation to their referents.

11   Having introduced an idea from Ceccato's 'operational analyses' into Piaget's model, I today believe that the motion necessary in perception need not be physical (of limbs or eyes), but can be replaced by the motion of the perceiver's focus of attention (see Glasersfeld, 1981a).

12   The title of this volume, as Leslie Smith (1981), one of the few conscientious interpreters of Piaget, pointed out, was mistranslated as 'The grasp of consciousness' and should have been rendered as the 'onset' or 'attainment of consciousness'.

13 I owe knowledge of Caramuel's work to my late friend Paolo Terzi, who immediately recognized the value of that seventeenth-century author's Latin treatise, when it was accidentally found in the library of Vigevano. Caramuel, a Spanish nobleman, called to the Vatican as architect, mathematician, and philosopher of science, was then exiled as bishop to that small Lombard city, because he had had several disagreements with the Holy See.

14 It may be helpful to remember that the first sentence in Kant's *Critique of pure reason* (1781) reads: 'Experience is undoubtedly the first product that our intelligence brings forth, by operating on the material of sensory impressions.'

Chapter 6

Constructing Agents: The Self and Others[1]

For some 2500 years the western world has manifested an overwhelming tendency to think of knowledge as the representation of a world outside and independent of the knower. The representation was supposed to reflect at least part of the world's structure and the principles according to which it works. Although the picture might not yet be quite perfect, it was thought to be perfectible in principle. As in the case of a portrait, the 'goodness' of a piece of knowledge was to be judged by how well it corresponded to the 'real' thing. For reasons laid out in the preceding chapters, this way of thinking is not viable from the constructivist point of view. But if one denies that knowledge must in some way correspond to an objective world, what should it be related to and what could give it its value?

This is a serious question, because if we were to say that there is no such relation, we should find ourselves caught in solipsism, according to which the mind, and the mind alone, creates the world. As an explanatory model the doctrine of solipsism is not very useful. In fact, it is not a model at all and it explains nothing. Solipsism is a metaphysical statement about the nature of the world and leaves to others the task of explaining how the individual sets about to create its world. If an autonomous 'will' is invoked (e.g., Schopenhauer, 1819), some powerful 'wild cards' have to be borrowed from mysticism to achieve a semblance of coherence. In practice, solipsism is refuted daily by the experience that the world is hardly ever what we would like it to be.

Constructivism, as I explained earlier, has nothing to say about what may or may not *exist*. It is intended as a theory of knowing, not as a theory of being. Nevertheless it does *not* maintain that we can successfully construct anything we might want. Two principles are crucial in this regard.

The first is that cognitive organisms do not acquire knowledge just for the fun of it. They develop attitudes towards their experience because they like certain parts of it and dislike others.

. . . [H]uman beings never remain passive but constantly pursue some aim or react to perturbations by active compensations consisting in regulations. It follows from this that every action proceeds from a need which is connected with the system as a whole and that values likewise dependent on the system as a whole are attached to every action and to every situation favourable or unfavourable to its execution. (Piaget, 1970c, p.38)

Consequently human actions become goal-directed in that they tend to repeat likeable experiences and to avoid the ones that are disliked. The way they attempt to achieve this, is by assuming that there must be regularities or, to put it more ambitiously, that there is some recognizable order in the experiential world. As the biologist Humberto Maturana said:

A living system, due to its circular organisation, is an inductive system and functions always in a predictive manner: what happened once will occur again. Its organisation (genetic and otherwise) is conservative and repeats only that which works. (Maturana, 1970a, p.39)

One kind of knowledge, then, is knowledge of what has worked in the past and can be expected to work again.

The second principle is that from the constructivist perspective, knowledge does not constitute a 'picture' of the world. It does not represent the world at all — it comprises action schemes, concepts, and thoughts, and it distinguishes the ones that are considered advantageous from those that are not. In other words, it pertains to the ways and means the cognizing subject has conceptually evolved in order to fit into the world as he or she experiences it.

It follows that what we ordinarily call 'facts' are not elements of an observer-independent world but elements of an observer's experience. As Vico noticed in 1710, the word *factum* is the past participle of the Latin word for 'to make'. This was one clue that led him to formulate the epistemological principle that human beings can know only what human beings themselves have made by putting together elements that were accessible to them.

This question of accessibility, it seems to me, is of crucial importance in any discussion of what is empirical and what is not. Going back to the beginnings of empiricism, we found that Locke proposed two different sources for the generation of ideas: on the one hand, the senses and, on the other, reflection. Kant then pulled the rug from under whatever had remained of realism after the British empiricists. By proposing that space and time should be considered characteristic forms (*Anschuungsformen*) of the human way of experiencing, rather than properties of the real world, he eliminated any possibility of envisaging or visualizing a world before it has gone through our experiential procedure. If we accept this view, we must also accept 'philosophically' something that we can check out 'empirically' for ourselves: we are incapable of seeing, touching, hearing, and, indeed, knowing anything that is not framed in space and/or time. Everything that we might want to call 'structure' depends on space and time. Hence, whatever 'ontic' reality might be like, it makes no sense to think of it as containing anything that we could recognize as structure.[2]

The Illusion of Encoded Information

At this point one might say, let's forget about the British empiricists and let's forget about Kant, maybe we can make a case for realism in spite of them. After all, there have been quite a few philosophers since Kant who implicitly, if not explicitly, tended towards realism. None, however, has found a satisfactory defence against the age-old attack of the sceptics. If we want to think of knowledge as a picture of reality, we would have to be reassured that we could come to have a realistic picture that shows things more or less as they really are. But a test that might give us this assurance is precisely what we cannot make.

This problem has recently cropped up in the context of information processing. In that school, too, knowledge is often discussed in representational terms. The cognitive organism, it is said, comes to form representations, and these representations 'encode' information that has been gleaned from reality. However, Bickhard and Richie (1983) have shown that this is an illusion. A code is an arrangement of semantic links between items that signify and items that are signified by them. In order to create such a meaningful connection, one must have access not only to the signs or symbols that one intends to use, but also to the items one wants them to signify or symbolize. Note that this is, in fact, a formulation in contemporary terms of the criterion Vico introduced to distinguish the mystics' metaphorical use of language from the rational. Because the presumed ontological reality, always remains on the other side of our experiential interface, the second condition (access to the code-symbols' meaning) cannot be fulfilled. Hence it is an unfortunate distortion when people say that the signals we receive through our senses are a 'code' that conveys information about reality.

A particularly striking problem for the view that the senses transmit information was unearthed by Heinz von Foerster from the work of the German nineteenth-century physiologist Johannes Müller. He aptly called it the 'Principle of Undifferentiated Encoding':

The response of a nerve cell does *not* encode the physical nature of the agents that caused its response. Encoded is only 'how much' at this point in my body, but not 'what'. (Foerster, 1981, p.293)

In other words, signals sent to the brain by neurones in your finger tips or toes, in your ear, or in the retina of your eye, are *qualitatively* all the same. They embody the intensity of the particular perturbation, but no information about the nature of its cause. The picture of a world containing visible, audible, tangible, etc., things, can be constructed only from relations an interpreter establishes between the signals, e.g., which arrive together and what sequences occur.

For this constructive activity, the role of attention is crucial. As I mentioned earlier, several experimental psychologists independently showed that subjects can freely move their focus of attention in the perceptual field without physically moving their eyes or their bodies (Köhler, 1951; Lashley, 1951; Pritchard *et al.*, 1960; Zinchenko and Vergiles, 1972). This startling finding is important for any model of mental construction. It eliminates the need for the traditional assumption that sensory signals come in preordained clumps and it frees the mind as originator of coordination and relations.

I want to make clear, however, that experimental results, no matter how compatible they may be with the constructivist model, do not make the model 'true'. The empirical findings of undifferentiated encoding and the mobility of attention, being themselves the constructs of observers, cannot serve as a logical argument to prove that the senses do not provide information about the structure of an objective external world. This impossibility springs from the sceptics' insight that human knowledge cannot be tested by a procedure that would again involves the mechanisms of human cognition.

Yet, since constructivism claims that knowing is the building of coherent networks by assembling conceptual structures and models that are mutually compatible, compatible empirical findings are always encouraging. In this case, moreover, they legitimize the demand that those who maintain that we *do* receive objective information through the senses, come up with a plausible model to explain *how* such a communicative transfer might work.

In turn, the constructivist theory has the obligation to provide a model capable of showing how it comes about that, in spite of informational closure, we seem to have a remarkably stable experiential reality in which we carry on our daily living.

The Reality of Experience

We formulate explanations, we make predictions, and we even manage to control certain events in the field of our experience which is the reality we live in. All this, and especially any attempt at management, involves what we call common sense and at times also scientific knowledge. The second is mostly held to be the more solid. We rely on it, and it allows us to do many quite marvellous things.

For epistemologists, therefore, it has become indispensable to look at the method scientists use to construct their knowledge. Contemporary philosophers of science are much divided on this topic and argue about how rationality and its role in the formation of knowledge should be defined (e.g., McMullin, 1988). From my point of view, it is more profitable to examine what scientists actually do. Humberto Maturana has produced a useful description of the procedure that is usually called 'the scientific method', and

I believe that there are few scientists who would not basically agree with it. Over the years Maturana has formulated his view with some variations and I am here giving my own summary.[3] The procedure is divided into four steps that are carried out when a phenomenon (an experience or sequence of experiences) is deemed to need explanation:

1 The conditions (constraints) under which the phenomenon is observed must be made explicit (so that the observation can be repeated).
2 A hypothetical mechanism is proposed that could serve as explanation of how the interesting or surprising aspects of the observed phenomenon may arise.
3 From the hypothetical mechanism a prediction is deduced, concerning an event that has not yet been observed.
4 The scientist then sets out to generate the conditions under which the mechanism should lead to the observation of the predicted event; and these conditions must again be made explicit.

Throughout the four steps, what matters is *experience*. Observing is a way of experiencing and, to be scientific, it must be regulated by certain constraints. The hypotheses (by means of which the observations are related) connect experiences, not things-in-themselves. The predictions, too, regard what we expect to experience, not events in some independent world beyond the experiential field.

Seen in this way, the scientific method does not refer to, nor does it need, the notion of ontological reality. It operates and produces its results in the experiential domain of observers. Scientific knowledge, then, provides more or less reliable ways of dealing with experiences, the only reality we know; and dealing with experiences means to be more or less successful in the pursuit of goals. Scientific knowledge, then, is deemed more reliable than commonsense knowledge, not because it is built up differently, but because the way in which it is built up is explicit and repeatable. Paul Valéry, surely one of the wisest men of our time, wrote: 'Science is the collection of recipes and procedures that work always', and he explained:

> Our faith [in it] rests entirely on the certainty of reproducing or seeing again a certain phenomenon by means of certain well defined acts. (Valéry, 1957, p.1253)

The value of scientific knowledge, thus, is not dependent on 'truth' in the philosopher's sense, but on 'viability'.

Unlike the notion of truth, which would require a match, i.e., shared points and features, of the picture and what it is intended to represent, the notion of viability (which refers to actions and ways of thinking) merely requires fit. This is a relation characterized by the absence of shared points, because they would be points of friction or collision (see Chapter 2).

The concept of viability, however, does imply that there are or will be obstacles and constraints that interfere with, and obstruct, the organism's way of attaining the chosen goals. It is certainly not the case that 'anything goes'.[4] It is always possible that an ontic reality manifests itself by impeding some of our actions and by thwarting some of our efforts. But even if this should be the case, this ontic reality would manifest itself only in failures of our acting and/or thinking, and we would have no way of describing it, except in terms of the actions and thoughts that turned out to be unsuccessful.

Analysis of Empirical Construction

With this approach to the problem of knowing, empiricism has come full circle, returning to its original intent to examine the world of experience. It began with the hope that the world of experience would sooner or later reveal something of an ontic world beyond it, a world of objective reality. This hope was not fulfilled. Thus, if we continue to investigate the world of experience, it must be in the spirit of Kant's 'transcendental enterprise', that is, with the intent to find out how we come to have the apparently stable world in which at a certain point in our development, we find ourselves living.

The expression 'the world in which we find ourselves living' is not intended to echo Heidegger's metaphysical notion of being 'thrown' into the world. Instead, it springs from the Piagetian idea that some of the concepts that determine the structure of our experiential world are constructed during the sensorimotor period, prior to the age of 2 years, when we are anything but aware of what we are building. As adults, therefore, as Spencer Brown so elegantly said:

> Our understanding of such a universe comes not from discovering its present appearance, but in remembering what we originally did to bring it about. (Spencer Brown, 1973, p.104)

What we ordinarily call reality is the domain of the relatively durable perceptual and conceptual structures which we manage to establish, use, and maintain in the flow of our actual experience. This experiential reality, no matter what epistemology we want to adopt, does not come to us in one piece. We build it up bit by bit in a succession of steps that, in retrospect, seem to form a succession of levels. Repetition is an indispensable factor in that development. Without repetition there would be no reason to claim that a given item has any permanence beyond the context of present experience. Only if we consider an experience to be the second instance of the self-same item we have experienced before, does the notion of permanent things arise. This creation of 'individual identity' has momentous consequences (see Chapter 4). If two experiences that we want to consider experiences of one and the same item do not immediately succeed one another, then we must provide a way for that item to survive. That is to say, we are obliged to think of that individual item as subsisting somewhere while we are attending to others in the flow of our experience. Thus we come to construct 'existence' as a condition or state of 'being' that takes place outside our experiential field; and the things that partake of this existence need space in which to be and time in order to perdure while our attention is elsewhere. In other words, by creating individual identities which we can repeat in our experience, we have created a fully furnished independent world that exists whether or not we experience its furniture.

The Question of Objectivity

If we do accept this way of thinking as a working hypothesis, we shall have to account for a difference in conceptual constructs which, even as constructivists, we would not like to miss: the difference between knowledge that we want to trust as though it were objective, and constructs that we consider to be questionable if not downright illusory. Needless to say, this constructivist 'objectivity' should be called by another name because it does not lie in, nor does it point to, a world of things-in-themselves. It lies wholly within the confines of the phenomenal. For reasons I shall presently explain, I have tentatively proposed the term 'intersubjective' for this highest, most reliable level of experiential reality (see Glasersfeld, 1986).

As the term implies, this uppermost level arises through the corroboration of other thinking and knowing subjects. This introduction of 'others' might seem to be in flat contradiction of the constructivist principle that all knowledge is subjective. However, the apparent contradiction will disappear if I am able to show that, although the others are the individual subject's construction, they can nevertheless provide a corroboration of that subject's experiential reality.

The model of how we construct 'others' is, in fact, an extension of a suggestion Kant made in the first edition of his *Critique of Pure Reason*:

> It is clear: If one conceives of another thinking subject, one necessarily imputes to that other the properties and capabilities by which one characterises oneself as subject. (Kant, 1781, p.223)

The creation of others in our likeness does not happen all at once. It begins quite harmlessly with the child imputing the capability of spontaneous movement to items in the experiential field that do not stay put. The moon, water flowing and forming eddies in a river, trees swaying in the wind, are thought to move, as children do in their own experience, of their own will. This stage is sometimes characterized as animism. It is followed by the imputation of visual and auditory senses to animals. The toddler who wants to catch a frog soon learns that he has to approach as quietly as possible and from behind. He concludes that the frog can hear and see. Later, there are innumerable situations with those other experiential objects called people, that lead the child to impute to them goal-directed behaviour, deliberate planning, feelings, and experiential learning. Finally, these others are considered more or less like oneself.

Corroboration by Others

Once this level of sophistication is reached, a great deal of time is spent explaining, predicting, and attempting to control these 'others'. That is to say, one now has populated one's experiential field with models of others who move, perceive, plan, think, feel, and even philosophize, others to whom one imputes the kinds of concepts, schemes, and rules one has oneself abstracted from experience.

At this point, these models are thought to have some of the knowledge we ourselves have found viable in our own dealings with experience. Thus, when we make a prediction about how one of these others will behave in a given situation, the prediction is based on a particular piece of knowledge which we have imputed to that other. If, then, the other does what we predicted, we may say that the piece of knowledge was found to be viable not only in our own sphere of actions but also in that of the other. This bestows a second order of viability to the knowledge and the reasoning we assumed the other to have and act on.

To appreciate the value of this kind of corroboration, it is crucial to remember that the individual's construction of other constructing agents is no more a free construction than that of the physical objects with which we furnish our experiential world. It is a construction that is continuously impeded and thus oriented, but not determined, by obstacles that function as constraints.

It is obvious that this second-order viability, of which we can say with some justification that it reaches beyond the field of our individual experience into that of others, must play an important part in the stabilization and solidification of our experiential reality. It helps to create that intersubjective level on which one is led to believe that concepts, schemes of action, goals, and ultimately feelings and emotions are shared by others and, therefore, more *real* than anything experienced only by oneself. It is the level on which one feels justified in speaking of 'confirmed facts', of 'society', 'social interaction', and 'common knowledge'.

The development of this intersubjective reality is sketched out clearly at the end of four dialogues that Alexander Bogdanov published in 1909. Bogdanov, who has now been recognized as a forerunner of cybernetics, was a truly universal thinker. He worked as a physician, experimental biologist, and philosopher of science and sociology. He argued vehemently with his friend Lenin, who published a condemnation of his philosophy but let him continue his work. Shortly after he died in a medical experiment he carried out on himself in 1928, Stalin banned Bogdanov's books and he was virtually forgotten. Some years ago, Vladimir Sadovsky, a Russian colleague, copied the dialogues for me from a volume he had kept hidden since his student days and said: 'We, too, have had a constructivist.'

When I finally managed to have them translated into a language I could read, I was amazed and delighted. These dialogues are an admirably concise and lucid presentation of the instrumentalist aspect of constructivism and they provide an explicitly social component to the generation of intersubjective reality.

Knowledge, Bogdanov says, functions as a tool. How good a tool is, or how much better it could be, comes out when a group of people work together at the same task. When no one can suggest a further improvement, the tool will be called 'truth' (Bogdanov, 1909, pp.30–3).

Many a detail is still to be explored in this social construction of our experiential truth and reality, but the notion of collaboration and the concerted efforts to reach a goal is probably the most powerful principle.[5] Other facets could be pieced together from isolated insights in the works of sociologists and social psychologists. But — and this is crucial from the constructivist point of view — when it comes to the invention and improvement of tools, Bogdanov adds:

> Whether genius or simple worker, in their respective cognitive and practical creativity they are always single human beings. (Bogdanov, 1909, p.33)

In other words, no analysis of social phenomena can be successful if it does not fully take into account that the mind that constructs viable concepts and schemes is under all circumstances an individual mind. Consequently, also 'others' and 'society' are concepts constructed by individuals on the basis of their own subjective experience.

To return to what I have called 'corroboration by others', one might think, that it is more easily and much more frequently achieved by linguistic communication. Although this is a general assumption, we all quite often find ourselves in situations that confound it. Others may be telling us (or we may believe) that they think as we do, but what they say or do shows us, as the interchange goes on, that this cannot be the case. Although the words they use are the same as ours, the network of concepts they seem to have in mind is incompatible with the one we have built up. Communication is a far more complex affair than it seems and I shall return to it in chapter 7.

The Elusive Self

At frequent intervals in the above text I have used the first person pronoun and much of what I said makes it clear that this first person is assumed to be a constructor of knowledge. Thus the question arises whether the active agent, the 'subject' that is supposed to reside in this first person, can spontaneously construct knowledge of him- or herself. It has often been said that it cannot, and that self-knowledge arises only from interaction with other persons.

One way of coming to grips with that question was broached by Descartes. His method of doubting everything that could be doubted, led him to the conclusion that the one thing that remained indubitable was the fact that it was he who was doing the doubting. He did not have to interact with others, he simply said to himself, *cogito ergo sum*, and concluded that, as long as he was thinking, he, the thinker, *existed* (see Chapter 2).

What did he mean by the word 'to exist'? He based his conclusion on 'thinking' and therefore could not have had in mind Berkeley's definition, which invokes perception. Descartes believed that space and time constitute an absolute, observer-independent frame of reference, and it is safe to assume that he shared the common-sense view that the expression 'to exist' means no less and no more than having a locus with specifiable coordinates in that framework. If we say that 'X exists', we ordinarily mean that at some point in time it is at some point in space.

After Kant, however, the situation gets more complicated. If space and time are no longer considered properties of the ontic world but 'ways of experiencing', we shall have to admit that there are some things to which, though they do have specifiable coordinates in an observer's subjective spatio-temporal frame of reference, we would not want to attribute existence, e.g., hallucinations, mirages, and, closer to home, mirror images and rainbows.

In any case, to fit the constructivist theory, Descartes' pronouncement 'I think, therefore I am' has to be reinterpreted as 'I am aware of thinking, therefore I am'. To my mind, it is precisely this awareness of what one is doing or experiencing that is the foundation of what we ordinarily call our *self*. It does not have to be thinking in any elevated sense. If you are becoming aware of tying your shoe laces, you also become aware of the fact that there is a you who is doing it.

Given that, as I said earlier, we have not even the beginnings of a model of consciousness or awareness, it may seem odd to insist that awareness lies at the very root of the concept of self as an entity. Yet, no attempt to construct a viable theory of knowing, can avoid acknowledging the fundamental mystery of awareness. As Wittgenstein put it:

> The I occurs in philosophy through the fact that the 'world is my world'. . . The philosophical I is not the man, not the human body or the human soul of which psychology treats, but the metaphysical subject, the limit — not a part of the world. (Wittgenstein, 1933, par.5.641)

But there are two aspects to the concept of self. In the realist view, the self we perceive, by being perceived, becomes an object under the command of a perceiving subject. Berger and Luckmann expressed this very neatly:

On the one hand, man *is* a body, in the same way that this may be said of every other animal organism. On the other hand, man *has* a body. That is, man experiences himself as an entity that is not identical with his body, but that, on the contrary, has that body at its disposal. In other words, man's experience of himself always hovers in a balance between being and having a body, a balance that must be redressed again and again. (Berger and Luckmann, 1967, p.50)

In the constructivist view, the self we conceive, as well as its body, are necessarily the product of that active agent that Wittgenstein called the 'I' that is not part of the world. Whatever the other-worldly part of the self builds up is gauged according to its viability in experience. Thus there is a rather straightforward way to approach the component of the self-concept that *is* part of the experiential world. Instead of asking what the self is in the philosopher's sense, one can ask how we experience our self. This does not concern the mysterious entity that does the experiencing, but focuses on the tangible structure, the body that is experienced as one's own. Such an investigation takes the mysterious self-conscious entity for granted and proceeds to examine how that entity comes to recognize itself both as agent and as percept distinguished from the rest of its experiential field.

The Notion of Environment

Perhaps the most serious obstacle that has impeded a viable analysis of the complex concept of self is the traditional assumption that the dichotomy between an experiencing subject and what it experiences is basically the same as the dichotomy between an organism and its environment. When we observe those items in our experiential field that we have come to call other organisms, we can of course speak of their 'environment'. But as I argued earlier the distinction between an organism and its environment can be made only by an observer. The environment of an observed organism, therefore, is the experiential field in which the observer has isolated that organism. The organism itself has no access to items outside of it or, as psychologists would say, to distal data.

If we assume that our picture of the world, the knowledge that constitutes our experiential reality, is constructed by us piece by piece on the basis of experience, then we must also assume that the picture/knowledge we have of our self must be constructed in a similar way. In other words, just as we construct a model of a world, externalize it, and then treat it as though its existence were independent of our doing, so we construct a model of the entity that we call our *self*, and externalize it so that it ends up as 'a thing among other things' (Piaget, 1937, pp.7 and 82).

This construction, obviously, has many steps and takes time to accomplish. Let us begin by asking what is being constructed.

The concept of self seems simple enough when we refer to it in an accustomed context and in ordinary language. As a rule, we have no problem with expressions such as 'I did it all myself' or 'Well, I can't help it, I am like that'. Even the rather peculiar expressions 'You are you' and 'I am I' do not seem as peculiar as Gertrude Stein's 'A rose is a rose is a rose'. What we apparently have in mind when we make statements of that kind, is the individual identity or continuity of a person. However, as soon as we attempt to analyse what precisely it is that constitutes the continuity of our *selves*, we run into difficulties. There is an impression of ambiguity: even the experiential self seems to have several aspects.

The Perceived Self

First of all, there is a self that is part of one's perceptual experience. In my visual field, for instance, I can easily distinguish my hand from the writing pad and table, and from the pencil it is holding. I have no doubt that the hand is part of me, while the pad, the table, and the pencil are not.

Second, if I move my eyes, tilt my head, or walk to the window, I can isolate my self as the locus of the perceptual (and other) experiences I am having. This self as the locus of experience appears to be an active agent rather than a passive entity. It *can*, in fact, move my eyes, tilt my head, change location — and it can also attend to one part of the visual or experiential field rather than to another. This active self can decide to look or not to look, to move or not to move, to hold the pencil or not to hold it. Within certain limits, it can even decide to experience or not to experience.[6]

Beyond these, there are still other aspects of the concept of self. There is, for instance, the social self. As experiencing subjects, we enter into specific relations to others, and as actors, we adopt specific patterns or roles that eventually come to be considered characteristic parts of what we call our selves. But I am here focusing on the construction of a conceptual core, and shall therefore disregard the social aspects of the self, because their main development seems to take place during adolescence when the basic self is already established.

The pages that follow, therefore, treat of the two main parts of the essential concept, the self as locus of experience and the self as perceptual entity.

With the construction of permanent objects, the cognitive subject crystallizes some of the repeatable items it has constructed and treats them as external and independent. Thus a distinction arises that covers much of the organism/environment distinction by creating a 'subjective' environment. The externalized permanent objects now 'exist' in an external world structured by the spatial and temporal relations that have been abstracted from the objects as they were experienced.

This externalization goes hand in hand with the development of the ability to re-present the objects when they are not actually available. However, the 'internal' and 'external' that result from this two-fold construction are not at all the same dichotomy that an observer makes between an observed organism and its environment. They are both conceptualizations of the subject, and the subject uses them to distinguish items re-presented from memory and items actually being constructed from sensorimotor material. The distinction is not unlike the one we are forced to make when we wake up with the vivid scene of a dream 'before our eyes' and realize that it was, indeed, a dream and is quite incompatible with whatever we can construct with the sensory material that is actually available at the moment.

Both the internal and the external, however, are experience, and the division between them, therefore, is between two types of experience and not a division between an experiencing subject and ready-made 'objective' objects that are waiting to be experienced by someone.

Insofar as we become aware of experiencing, both types can provide the assurance that we are the subject that does it. When we wake up and realize that the sunny beach was part of a dream and that it is actually raining outside, we have no doubt that it was no less we, ourselves, who were dreaming than it is we who are now looking out through the window. Yet, neither gives us a view of ourselves 'as a thing among other things'.

Sensory Clues

How, then, do we come to 'see' our self? How do we come to know that the hand holding the pencil is *our* hand? It is a long story and all I can do is point out a few of the steps that would seem essential.

It probably begins with the infant's discovery that, having noticed moving shapes in its visual field, there is a way to distinguish some of them. When, for example, the mother's hand moves across the infant's visual field, what the infant experiences is purely visual. However, when the infant's own hand moves across its visual field, the visual experience has the possible corollary of a kinaesthetic experience, namely the sensory signals the infant gets from the muscles that happen to be involved in generating the hand's movement. A little later, the difference between the two experiences is significantly increased by the realization that the hand's movement can be reliably initiated at will, whereas the movement of the mother's hand cannot.

When infants touch some part of their body with their own hand, tactual signals are generated on both sides of the point of contact. This makes possible a reliable distinction between touching oneself and touching or being touched by other things (when tactual signals are generated only on one side of the contact). This distinction is surely made by every kitten that plays with its litter mates and discovers that biting its own tail is different from biting someone else's. There is no question that they quickly learn to distinguish their own tail when it comes to biting. Although in the kittens' case there is probably no reflective abstraction, the experience nevertheless generates for them a notion of self that has practical implications for their future behaviour.

125

Analogous experiences enable the child eventually to know whose hand is holding the pencil.

Reflected Images

Once children have coordinated tactual and proprioceptive elements to form some notion of their own body and when visual recognition of their own limbs has become reliable, this sets the stage for a considerably more complex experience of the physical self: the recognition of one's own shadow and one's image in a looking glass.

Gordon Gallup (1977), in a survey of research on self-recognition in primates, came to the conclusion that only the great apes have the ability to recognize their mirror image as their own. Monkeys quickly learn to discriminate their shadows, reflections, and mirror images from other moving objects, but do not appear to relate them in any way to themselves. For a long time they seem uncertain whether it is or is not another animal.[7]

The simple synchrony of movement between, say, a paw and its shadow or reflection does not seem sufficient to establish the link to the self. It may be that a causal connection must be constructed from a deliberate motion to its reflected counterpart, and that it is this connection which differentiates the motion of a mirror image from the motion of another object or organism.

At 2½ or 3 years of age, children will relate a coloured patch they see on the image in the mirror to the patch that has been surreptitiously placed on their own forehead, but nevertheless most of them will still go to look behind the mirror. They have apparently related the image to themselves, but the self has not yet established its unique position in space (see Zazzo, 1979, p.241).

The child who stands in front of a looking glass, sticks out his tongue, and contorts his face into all sorts of grimaces, gets constant confirmation of a causal link. The mirror image is as obedient as the subject's own body and as completely under the subject's control. It can thus be integrated with the body percept, expanding it by providing visual access to otherwise invisible aspects. And like the body image it is a visual percept, a sensorimotor self that is being experienced. It tells us nothing about the self as the agent of experience.

The Social Self

Much more complex than these very basic considerations would be the analysis of the social component in the construction and evolution of an individual's concept of self. As the social psychologist Paul Secord explained:

> Perhaps most important to his developing idea of a person as a somewhat stable entity in his world is his realisation that other persons

> behave in predictable ways . . . Only with time and much experience does the individual eventually identify at least some properties of a relatively stable nature associated with himself. Both self as object and self as agent are relevant here. (Secord and Peevers, 1974, p.121)

From the constructivist perspective, this may well be a viable view. I do not know whether it has ever been corroborated in the observation of children. If it has not, it would be a be a rewarding investigation for social psychologists.

There is, however, a fundamental theoretical complication. If it is others from whose reactions I derive some indication as to the properties I can ascribe to myself, and if my knowledge of these others is the result of my own construction, there is an inherent circularity in that procedure. In my view, this is not a vicious circle, because we are not free to construct others in any way we like. As with all other constructs, the 'models' we build up of others either turn out to be viable in our experience, or they do not and have to be discarded.

This dependence on viability in our construction of other individuals has a consequence that leads into the direction of ethics, a realm that is no less opaque for constructivism than for other rational theories of knowledge. Nevertheless, the fact that the individual needs the corroboration of others to establish the intersubjective viability of ways of thinking and acting, entails a concern for others as autonomous constructors. If we force them in any way to conform to our ideas, we *ipso facto* invalidate them as corroborators.

In fact, this is another formulation of Kant's 'practical imperative':

> Act always in such a way that the humanity, in your own as well as in other persons, is treated as end and not just as means. (Kant, 1785, p.429)

Thinking beings, he explains, are ends in themselves and no other purpose must be substituted for this (ibid., p.428). Strictly speaking, this is not an 'ethical' precept but a prerequisite of ethics. It simply asserts that we have to consider other people's humanity and that we ought not to treat them as objects. All philosophy of ethics is implicitly based on this assumption. Yet, it does not say why it should be so. Constructivism provides at least one basic reason. From its perspective, the concern for others can be grounded in the individual subject's *need* for other people in order to establish an intersubjective viability of ways of thinking and acting. Others have to be considered because they are irreplaceable in the construction of a more solid experiential reality. This in itself does not constitute an ethical precept either, but it does supply a rational basis for the development of ethics. Let me emphasize that ethics itself cannot actually be based on the viability of schemes of action or thought, because this viability is always gauged in the context of specific goals — and it is in the choice of goals that ethics must manifest itself.

Conclusion

Empirical facts, from the constructivist perspective, are constructs based on regularities in a subject's experience. They are viable if they maintain their usefulness and serve their purposes in the pursuit of goals.

In the course of organizing and systematizing experience, the subject creates not only objects to which independent existence is attributed but also *others* to whom the subject imputes such status and capabilities as are conceivable, given his or her own experience.

Where knowledge is concerned, the concepts, theories, beliefs, and other abstract structures which the individual subject has found to be viable, gain a higher degree of viability when successful predictions can be made by imputing the use of this knowledge to others. The additional viability can be interpreted as indicating intersubjectivity and constitutes the constructivist substitute for objectivity. This implies that the individual has a need to construct others and to keep these models of others as viable as possible because only viable others can lend the highest level of support to the subject's experiential reality.

As to the concept of self, constructivism — as an empirical epistemology — can provide a more or less viable model for the construction of the experiential self; but the self as the operative agent of construction, the self as the locus of subjective awareness, seems to be a metaphysical assumption and lies outside the domain of empirical construction.

Notes

1 Some of the ideas presented in this chapter were first published in Glasersfeld, 1979 and 1989b.
2 This is one root of the difficulty we encounter in the mystics' metaphorical use of words such as 'oneness'; the word is associated with a concept the make-up of which involves separation from a background and the constitution of a bounded unit; whereas the mystic's notion should have no background because it is intended to be infinite and all-comprehensive.
3 A recent version can be found in Maturana, 1988, pp.34–5.
4 When Paul Feyerabend (1975, p.23) used this phrase, the context made quite clear that he did not intend, as some of his critics imputed to him, anything at all, but rather anything that seems useful.
5 Pietro Barbetta, recently sent me his Italian translation of Piaget's *Études sociologiques* (1965), a work to which I had never had access before. There I discovered that Piaget approached the social construction of knowledge in the very same way. Fortunately, Leslie Smith is now in the process of publishing an English translation of these important Piagetian essays.
6 While Oriental philosophy has always cultivated this autonomy of the experiencer, the western world, in defence of its traditional belief in an objective reality, has tended to consider experience as obligatory, inevitable, and rather passive.
7 Many of us may have observed the same with cats and dogs in our homes.

Chapter 7

On Language, Meaning, and Communication

Sometimes, when describing a scene or an event, it happens that one takes back a word and replaces it with another. The first one somehow did not seem to fit. There was an uneasiness, a slight perturbation, and this triggered the search for a more satisfactory expression. You may have noticed it in speaking, but more often probably in writing (all those hand-written congratulations or condolences that one had to write again because a single word seemed inappropriate!).

The psychologists who categorize the use of language as 'verbal behaviour' do not seem to have taken this phenomenon into account. Had they considered it, they might have noticed that these instances of self-correction cannot be explained in terms of environmental stimuli. What causes them is something *inside* the speaker or writer, a kind of monitoring that checks the linguistic formulation for its suitability in view of an intended effect. In my view, this is not a negligible feature.

As a rule, the use of language is purposive (see Glasersfeld, 1976a). There are, of course, occasions when we use a word or two blindly, without thinking — for instance, when our hammer hits the thumb instead of the nail we are driving into the wall — but then it is mostly bad language and not addressed to anyone in particular. On the whole, people speak with a specific intention. It may be a story they want to tell, an instruction they want to give, or simply to describe something they have seen or felt. In all these cases, the speakers have the re-presentation of a more or less detailed conceptual structure in their head. The words they utter and the sentences they form are those which, at least at that moment, seem to fit the story the speaker wants to tell. However, even the simplest events are not quite the same in the experience of different people. The process of associating words with sensorimotor experiences and the concepts abstracted from those experiences, is a subjective affair. Communication, therefore, is not a straightforward exchange of fixed meanings. Indeed, before one can establish the meaning of words, phrases, and sentences, there is the notorious problem of what we mean when we speak of meaning.

The Semantic Basis

It was de Saussure who, in the first decade of this century, postulated that the 'semantic' connection, the link between words and meanings, is based on the associational link between 'sound-images' and 'concepts'. Sound-images, as he emphasized more than once, are not to be confused with the physical sound of spoken words. They are abstractions from the auditory experiences of the sounds, just as, for instance, the concept of 'apple' is an abstraction from apple-experiences. Once a sound-image is linked to a concept, the combination constitutes a 'linguistic sign', which is 'a two-sided psychological entity' (de Saussure, 1959, p.66). A simple diagram can show the connections:

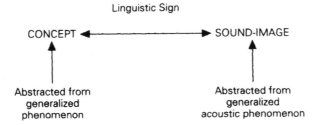

Figure 7.1:   The Semantic Connection (after de Saussure)

It is this psychological two-sidedness that makes possible what I have called the 'symbolic' use of words and language. The moment an experiential situation can be assimilated to an existing concept by the speaker of a language, it calls up the sound-image associated with it; and vice versa, the moment a heard speech sound can be assimilated to a sound-image it calls up the associated concept. It is a two-way connection and functions in both directions. Because this is not a behavioural but a mental affair, I call it 'symbolic' (Glasersfeld, 1974). It is fundamentally different from the use of words and other signs in 'signalling'.

I have earlier mentioned the example of a dog who obediently sits down whenever his mistress utters the word 'sit'. To follow the command, the dog must have something like a sound-image in order to isolate the sound 'sit' from all noises in his auditory field. For him this sound-image is not associated with a concept of sitting down, but simply with the behavioural response which he has been trained to produce whenever this particular sound is uttered by his mistress. It is the signal for an action, not a symbol that evokes a concept or a mental re-presentation.[1] Besides, a signal is constituted by a one-way connection: when the dog happens to sit down spontaneously, this does not call up in him the sound-image of the command and he could not describe what he is doing by uttering 'sit'.

Among humans, the context that exemplifies signalling and practically excludes symbolic interaction is the military. Commands are given to be executed, not to call up concepts, let alone thoughts, in the receivers. Since the commands frequently have the form but not the function of linguistic signs, this use of words is adequately described as verbal behaviour. It has little in common with the ordinary, conceptual use of words and language.

In my work on conceptual analysis I have found de Saussure's basic insight extremely helpful. To show its central role in the constructivist model of language, I have incorporated it in a more comprehensive diagram.

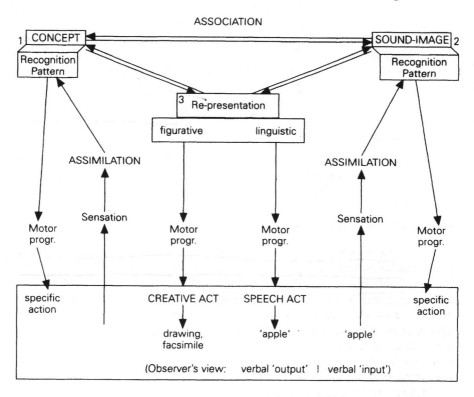

Figure 7.2: The Role of Concepts and Re-presentations in Linguistic Interactions

1   The distinction between 'concept' and 'recognition pattern' is developmental and functional. The recognition pattern is formed earlier and makes possible the coordination of specific actions with particular percepts. This underlies all phenomena that can be adequately described by the behaviourist notion of stimulus–response mechanisms. For example, when you cross the street and are one step away from the curb, you raise your foot higher than in the preceding steps. You do not have to become conceptually aware of the curb, your perceptual system recognizes the sensory pattern and this triggers the motor programme that has proved viable in the past. In Piagetian terms, the sequence is your 'curb negotiating scheme'. Much of our daily actions

function in this way and require no conceptual involvement — but this does not justify the behaviourists' assumption that concepts do not exist. The recognition pattern merges into a concept when it can be called up, spontaneously or by a word, in the absence of the sensory signals that characterize it and without triggering an action.

2   The distinction between 'sound–Image' and 'recognition pattern' is analogous. The recognition pattern that allows you to recognize particular auditory signals as a word of the language is formed earlier than the representation of the sound-image that allows you to produce the word in a speech act. This, as I have mentioned before, is captured by the linguists' distinction of passive and active vocabulary. The configuration of the right side of the diagram is the mirror image of the left, because there is again the stimulus–response pathway that results from training or practice and underlies the power of commands. There is no difference between the elements of the two pathways, because the fact that, on the right, the trigger is an auditory experience of phonemes (sounds characterized as linguistic) is irrelevant. The specific action that has been linked to it as response could be triggered just as well by a gesture, a light signal, or a flag, if these items had been included in the training.

3   The main addition to the Saussurian break-down is the introduction of 're-presentations'. In my view, this addition is essential, because the ability to call up re-presentations in listeners or readers is what gives language its enormous power and differentiates it from all forms of signalling. It is also the feature which distinguishes my model of language and communication from those of Humberto Maturana and Richard Rorty, many of whose other ideas are perfectly compatible with radical constructivism.

The circle of two-way connections linking concepts, sound-images, and re-presentations makes possible that each can be called up by both others. If I now present you with a word that is totally out of context, e.g.,

rhinoceros

you may not know what to do with it, but if I asked you what it *means*, you could certainly give an answer. You might say that it is a dangerous beast, has a notorious horn, and lives in Africa. You could say this, because the word called forth a visual re-presentation formed from pictures you have seen, from an experience on a safari, or from a visit to the zoo. It might also call forth all sorts of other things — a play by Ionesco, a particularly boorish person at your office, or whatever else you happen to have associated with the word. This wider range of associations (sometimes referred to as 'connotation') is what poetry relies on. Here, however, I am interested in the primary association, between the word and the animal as an experience.[2]

There may be words, and there are certainly names, that call forth a re-presentation that was not formed from actual experience but vicariously from a description. Most people will have constructed some image of Arcadia even if they have not seen it painted, or of Mme Bovary or Sir Galahad though they have never seen them in a movie.

Language Games

The point that words do call up re-presentations is important in view of the present fashion to consider language a social phenomenon that is simply a particular form of interaction. As far as I can see, this fashion sprang from a somewhat simplistic interpretation of statements found in Wittgenstein's later work, the *Philosophical Investigations* of 1953. There, in the first five pages, he neatly separates the 'signalling' function of words from the 'symbolic' one that relies on the generation of re-presentations. He does not use the two terms I have chosen, but his examples show that he is making a similar distinction. To characterize the first function, he describes a builder who calls to his assistant: 'block', 'pillar', 'slab', 'beam', according to what he needs at the moment. All that matters is that the assistant responds by bringing the right thing.

Of the second function Wittgenstein says that it is like 'hitting the keys of a *Vorstellungsklavier*, i.e., a kind of piano that produces re-presentations (1953, par.6). He then proceeds to complicate the builder/assistant situation by including colours, numbers, and eventually questions in the builder's utterances. This is to demonstrate that in the ordinary practice of language the two functions are interwoven in different ways that depend on the context. He described this variety as a collection of language games.

The expression 'language game' has become a catch-word. Together with Wittgenstein's assertion that the meaning of a word is the way it is used, it has misled many to think of language as an objective entity that comes to be shared by all members of a society.

In my view, this is a misinterpretation. It fails to take into account that Wittgenstein made both these points as part of his polemic against Frege's theory of reference and some of the things he himself had written in his *Tractatus* some thirty years earlier. In the first part of his later book, he was mainly concerned with demolishing the notion that words refer to observer-independent objects, and he did it brilliantly by drawing attention to the difference between speaking of a 'chair' and speaking of the 'shape' of a chair (ibid., par.35). The chair can be pointed to, he says, but the shape cannot.

There are, indeed, many words in a language, the meaning of which cannot be explained by simply pointing to perceivable referents. Shape is a good example, because one cannot see it the way one sees a colour or feel it the way one feels heat or cold. The meaning of the word 'shape', therefore, is not something independent that the senses can pick up. With this example,

Wittgenstein quite successfully refuted the logicians' theory of reference. But he did not answer the question *how* a meaning for the word 'shape' could be generated.

The Construction of Meaning

From the constructivist perspective, to grasp a shape requires following a perceived outline (e.g., a border formed by a difference of colour or texture) with the movement of a finger or visual attention, and then abstracting a pattern from the movement. In short, shape is an action pattern carried out by a perceiver.

Throughout his work, Wittgenstein was searching for an answer to the question he had raised early in his *Philosophical Investigations*: 'What is the relation between name and thing named?' (par.37). More than a hundred pages later he asks:

> Every sign by *itself* seems dead. *What* gives it life? — In use it is *alive*. Is life breathed into it there? — Or is the *use* its life? (Wittgenstein, 1953; par.432, p.128ᵉ)

The notions of language game and meaning-as-use provide a perfectly viable description of linguistic interactions, but they do not explain how the individual language user becomes a proficient player.

Wittgenstein was, of course, well aware that one could think of 'use' as individual and private, consisting in a person's calling up associated experiences. He had mentioned this long before in his notes for students, but he added that there was something occult about this mental capability and that it should therefore be avoided. He hoped that it could be avoided by assuming that the meaning of a linguistic expression could be captured by observing the way a social group uses it in their language games (1958, pp.3–5).

This, I think, was an illusion. Wittgenstein, who had undoubtedly one of the sharpest intellects in our century, struggled until his death to convert his notion of meaning and truth into a logical certainty, but the final pages of his last notebook (1969) show that he did not succeed in eliminating the subjective element.

The subjective element is inevitable because the semantic connection that ties sound-images to meanings has to be actively formed by each individual speaker.

> . . . a speech-sound localised in the brain, even when associated with the particular movements of the 'speech organs' that are required to produce it, is very far from being an element of language. It must be further associated with some element or group of elements of *experience*, say a visual image or a class of visual images or a feeling

of relation, before it has even rudimentary linguistic significance. (Sapir, 1921, p.10)

When, a couple of pages ago, you read the word 'rhinoceros', you had no idea what game I was playing nor what use I was making of the word. Yet, you produced *your* re-presentation. I emphasize that it was yours, because it was you who had at some earlier point in time extracted or abstracted it from your own experience. It was this re-presentation which, at that moment, 'breathed life' into the word for you.

There is no question that your association between the word and whatever experiential elements it called forth in you was formed because of the way you heard others use the word. But this was not (and is never) a simple transmittal. No one but you can make *your* associations, and no one but you can isolate *your* sound-image and whatever *you* conceptualize in your experiential field. The reason why you isolated and associated the two items may have been as simple as someone pointing at a picture of the animal in a book and uttering speech sounds that were new to you; or your safari guide may have said 'Ah, a rhinoceros', as the horned creature broke out of a thicket into the peaceful landscape in front of you.

In other constructions of meaning, say, in forming the association of the abstract concept and the word 'number', it would require several steps of reflection upon your own mental operations (see Chapter 9). In all cases, an experiential context is needed, as well as a speaker to produce the language sounds that allow you to construct a specific sound-image.

Thus, there is indeed a necessary social component in the formation of semantic connections between words on the one hand, and concepts and re-presentations on the other. Once the connection is formed, however, the word points to and brings forth nothing but the particular re-presentational material the language user has associated with it in his or her mind. Hence it is their use that brings words to life, but to use them, one does not need others or a social context — one uses them whenever one wants to in one's own thinking, dreaming, and speaking.

In the presentation of some of Vico's ideas in Chapter 2, I briefly mentioned that he made a useful distinction between the common and the metaphorical use of words. In the common use, words are intended to point to elements of experience and concepts abstracted from experience. In the metaphorical use, they are intended to point beyond experience to a world of imagination. The latter is the mode of the poet and the mystic. Bernard Shaw described it splendidly in a few lines of cross examination in his *St Joan*:

JOAN: . . . you must not talk to me about my voices.
ROBERT (inquisitor): How do you mean? voices?
JOAN: I hear voices telling me what to do. They come from God.
ROBERT: They come from your imagination.
JOAN: Of course. That is how the messages of God come to us.
    (Shaw, *St. Joan* 1923, Scene I)

Language and Reality

Maturana has frequently defined the phenomenon of language as 'the consensual coordination of coordinations of action' (e.g., 1988; pp.46–47). I can interpret this by saying: to enter into the use of a language I have to coordinate sound-images and re-presentations of experiences in such a way that the pairs I construct seem compatible (i.e., are in coordination) with the pairs other speakers of the language have constructed. For Maturana, too, some of the actions are the uttering of speech sounds, but what is coordinated with them are not acts of representation but other actions. 'Words', he says, 'are not symbolic entities, nor do they connote or denote independent objects' (ibid., p.47). I certainly agree that words do not connote, denote, or refer to independent objects, but in my model they have specific connections to parts of individual experience. It seems that Maturana discards the possibility of words calling forth re-presentations because in his view there is only representation without the hyphen, which is the sort that entails the illusion that one could have an image of reality.

Rorty takes a similar position when he says that one should not 'view language as a medium for either expression or representation' (1989, p.11). For him it follows from the insight that reference to real objects and true representations of reality are useless notions. Again, I agree with his evaluation of the philosophers' notions of reference to real objects and their representations; but I see no reason why a model of language should not draw on the human ability to recall past experiences and to link them with sound-images. The unexpected interlude with the rhinoceros has, I hope demonstrated that you all possess that ability. Without it, I claim, you would never benefit from linguistic interactions with others, nor could you learn anything from books.

The frequent objection against the use of re-presentation in a functional model of language seem to spring from a root analogous to that of some objections against the theory of knowledge presented by Kant. In Kant's case there was the mistranslation of his term *Vorstellung* which, as I explained earlier, intends an individual creation, not the reproduction of an objective original. In the approach to language the misinterpretation stems from an erroneous notion that is widespread and considered plausible among language users and those who begin the think about how language could work.

By the time human beings are 6 or 7-years old, they have developed a considerable mastery of the language spoken in the social group in which they grow up. They can use words and be understood by others and they understand a great deal of what others are saying. They are not yet at an age where they ponder *how* such understanding might be possible. Nor do they have reason to suspect that the things with which they have associated words are elements of their own experience rather than things that exist in themselves in an environment that is the same for everyone. Hence it seems quite natural that words should refer to independent objects and that their meaning, therefore, be universal, in that it is 'shared' by all individual speakers. Every day, these apparent facts are confirmed innumerable times, and if at some later stage reflections about language are entertained, they will almost inevitably be grounded in this conviction as a premise. It seems as unquestionable as that the sun moves round the earth, but is even less propitious as the basis of an explanatory model.

From the constructivist point of view, the notion of 'sharing' (see Chapter 2) does not imply sameness but compatibility in the context of mental constructs. Every learner of a language must construct his or her word meanings out of elements of individual experience and then adapt these meanings by trial, error, and hanging on to what seems to work in the linguistic interactions with others. There is no doubt that these subjective meanings get modified, honed, and adapted throughout their use in the course of social interactions. But this adaptation does not and cannot change the fact that the material an individual's meanings are composed of can be taken only from that individual's own subjective experience. For this reason, careful investigators of social interaction follow Paul Cobb and speak of meaning as taken-to-be-shared, which does not imply actual sameness (see Cobb, 1989).

Language, then, opens a not quite transparent window on the abstractions and re-presentations individual speakers glean from their experiential reality, but it does not, as analytical philosophers were hoping, open any window on the ontological reality of an independent world.

It may be useful to repeat that this is not a denial of reality, nor does it deny that we interact with other speakers and with an environment; but it does deny that the human knower can come to know reality in the ontological sense. The reason for this denial is simply that the human knower's interactions with the ontic world may reveal to some extent what the human knower can do — the space in which the human knower can move —, but they cannot reveal the nature of the constraints within which the human knower's movements are confined. Constructivism, thus, does not say there is no world and no other people, it merely holds that insofar as we know them, both the world and the others are models that we ourselves construct.

In the construction of the models that constitute a large part of our knowledge, language is of course an important tool. It serves in many ways and one of the most powerful is that it can provide instruction for experiences that one has not yet had. Let us assume you are reading a novel in which the heroine at some point travels to Paris. One morning a friend picks her up at her hotel in the Latin Quarter and, though their conversation while they are walking is about art, he now and then draws her attention to things they are passing — we are crossing the Pont Saint Michel; now, if we turned right here, we would come to the Bastille, but we have to turn left to get to the Louvre.

A few weeks later, you happen to be in Paris for the first time, and if you remember the bit from the novel, you have a piece of the map of Paris in your head. It is a minute piece and you may never have occasion to use it, but that is beside the point. You were able to build it up from the text, because you could assimilate what you read to your knowledge of bridges and what it means to turn right or left in city streets. This is the way you have built up, through linguistic communication, a vast number of models that you could then use in your actual experiential reality.

Theory of Communication

That compatibility is a more adequate concept than the sharing of meaning, emerges also from the scientific investigations of communication that were begun in this century. Claude Shannon's (1948) work on communication was revolutionary because it established incontrovertibly that the physical signals that pass between persons in communication (if they use language, the sounds of speech or the visual patterns of print or writing) do not carry what is ordinarily considered the meaning. Instead, they carry instructions to select particular meanings from a list, which, together with the list of convened signals, constitutes the communication code. This two-fold list is a 'code', and it constitutes the framework of information within which the signals function.

> Information is a quantity of selection. The nature of the entities selected, like the issue of 'meaning', does not enter into the theory. (Pask, 1961, p.124)

If the two lists of the code are not available to a receiver before the linguistic interaction takes place, the signals have no information for that receiver and he can construct no meaning.

To give a simple example, if you ask at the information counter at an airport at what time the plane from Boston is scheduled to arrive, you may get the answer '2.45 pm'. The string of acoustic signals that constitutes this utterance can have no meaning for you unless you have a particular conceptual schema in your head. This schema is part of the present-day English code that divides the day into twice twelve hours and each hour into sixty minutes. As a competent speaker of English, you are aware of that schema, and the received signals enable you to select one particular point of the 1440 possible points that the conventional temporal schema contains. The particular selection then acquires meaning for you in relation to the purpose that led you to ask the question.

A more complex example may throw light on the role of language. In the days when telephoning across the Atlantic was still arduous and expensive, I occasionally had to send a telegram to an Italian friend of mine in Austria. I lived in Georgia, in the South of the United States, and no one at the telegraph office there spoke Italian. Of course this did not matter, because I wrote out my message in block letters and the telegraphist typed them one by one into his machine. He did not have to know what the message said, he only had to know Morse code. At the Austrian end of the line, the string of Morse code signals was decoded into letters by a telegraphist who did not know Italian either, and the result was sent to my friend. It was he who recognized words, connected them to form sentences, and converted the message into conceptual structures in his head. Hence there were two codes involved: the Morse code of the telegraphists and the 'code' of the Italian language. However, there is a big difference between the two. The Morse code is an internationally agreed on list that has two columns, giving the letters of the alphabet in the one and, in the other, dots and dashes in various combinations. All one has to do to turn into an amateur telegraphist is to acquire a copy of the list that shows the Morse code. No such list exists for Italian or any other language. Although dictionaries do contain a more or less complete list of the words available in a language, they present the words' meanings by using other words. A dictionary may help to expand the range of one's linguistic code, but one cannot begin to learn from it *what* the words encode. Moreover, words by themselves constitute only a part of linguistic meaning. There is the other level of meaning that arises from their combinations. This other level is called 'syntax' and it is described in grammar books — and this description is again in words.

To speak of language as a 'code' and of interpreting it as 'decoding' is useful up to a point, but it tends to obscure a second level of interpretation that is of a different nature, yet in most cases equally indispensable. If the context for which a piece of language was intended is unknown, one may be able to decode it (if it is in a language one knows) and still be at loss as to an interpretation. Taken out of the context of the information counter at the airport, the utterance '2.45 pm' can be decoded and yields the same point in the daily time scale, but there is nothing one could *do* with it. There is no way of fitting it into a larger network of concepts and intentions. We shall return to this problem later in the chapter.

How We May Come to Use Language

When you learned to decode the sounds of your first language, you did not do it with the help of dictionaries and grammar books. Chomsky and his followers hold that, because the results of language acquisition seem 'instantaneous' (1986), children must have an innate device for this purpose. Anyone who has methodically followed a child through the early phases of language acquisition has noticed that there is no instantaneous advance. The steps are tentative and provisional and meanings are continuously expanded and reduced (see Tomasello, 1992). Besides, in order to substantiate the claim that the human animal has a specific genetically determined device for language, it would be necessary to show why and how such a genetic oddity could have been produced by natural selection.

In fact, it is not at all surprising that the child's construction of word meanings and grammatical structures is always tentative and subject to experimentation. If it is the case that meanings or concepts in general are not inherent in words and therefore cannot be physically transported from one person to another, the only possible answer seems to be the one Wittgenstein suggested: children make the semantic connections between words and concepts by observing the language games the adults around them are playing. But 'observing' is a rather misleading understatement in this context.

Although it is often said that normal children acquire their language without noticeable effort, a closer examination shows that the process involved is far from simple. Most of it children achieve on their own by experimenting with utterances, changing them, and experimenting again in their interactions with others. Occasionally a parent sets out to teach them the meaning of a word. These attempts bring out quite clearly what the difficulties are.

If you want your infant to learn, let us say, the word 'cup', you will go through a routine that parents have used from time immemorial. You will point to, and then probably pick up and move an object that satisfies your definition of 'cup'. At the same time you will repeatedly utter the word. It is likely that mothers and fathers do this intuitively, without any particular theoretical basis. They do it because it usually works. The reason why it works is not difficult to see. There are at least three essential steps the child has to make.

The first consists in focusing attention on some specific sensory signals in the manifold of sensory signals which, at every moment, are available; the parent's pointing provides direction for this act, even though at first, what is being pointed to, is usually not clear or ambiguous from the child's point of view.

The second step consists in isolating and coordinating a group of these sensory signals to form a more or less unitary item or 'thing'. The parent's moving the cup is a help in this task because it accentuates the relevant figure as opposed to the rest of the visual field that is to form the ground.[3]

The third step, then, is to associate the isolated visual pattern with the auditory experience produced by the parent's utterances of the word 'cup'. Again, the child must first isolate the sensory signals that constitute this auditory experience from the background consisting of the manifold auditory signals that are available at the moment; once more, the parent's repetition of the word obviously enhances the process of isolating the auditory pattern as well as its association with the unitary visual item.

If this sequence of steps provides an adequate analysis of the initial acquisition of the meaning of the world 'cup', it is clear that the child's meaning of that word is made up exclusively of elements which the child abstracts from her own experience. Indeed, anyone who has methodically watched children acquire the use of new words, will have noticed that what they at first isolate as meanings from their experience is often only partially compatible with the meanings the adult speakers of the language take for granted. Thus the child's concept of cup often includes for quite some time the activity of drinking and sometimes even the milk that happened to be in the cup.

It may take many thwarted linguistic interactions and repeated cropping and adding of the child's meaning before the re-presentation associated with the word 'cup' has been accommodated to fit, more or less at least, the many ways in which the word is used by the speakers of the language. As adults, we tend to forget how much groping, guessing, and modifying was needed before we ourselves constructed a meaning of 'cup' that was viable in contexts as divergent as china cabinets, soccer championships, golf greens, and the anatomy of hips and shoulders. In fact, the process of tuning and accommodating the meaning of words and linguistic expressions continues for each of us throughout our lives. No matter how long we have spoken a language, there will still be occasions when we realize that we have been using a word in a way that turns out to be idiosyncratic in some particular respect.

To Understand Understanding

Once we have come to see this essential and inevitable subjectivity in the construction of linguistic meaning, it is no longer possible to maintain the preconceived notion that words convey ideas or knowledge and that the listener who understands what we say must necessarily have conceptual structures that are identical with ours. Instead, we come to realize that understanding is always a matter of fit rather than match. The concept of viability, intended as a function of fitting into an experiential context, is as useful in the domain of linguistic communication as in the theory of evolution and epistemology.

The receiver of a piece of language, be it a word, a sentence, or a text, faces a task of interpretation. A piece of language directs the receiver to build up a conceptual structure, but there is no direct transmission of the meaning the speaker or writer intended. The only building blocks available to the interpreter are his or her own subjective conceptualizations and re-presentations.

> To understand another's speech, it is not sufficient to understand his words — we must understand his thought. But even that is not enough — we must also know its motivation. (Vygotsky, 1962, p.151)

Social constructivists, who claim Vygotsky as their founding father and now and then argue quite vehemently against radical constructivism, may be surprised at this quotation. It is fully compatible with my view. If there is a difference, it resides in the explanation of how the incipient language user develops this understanding. In Vygotsky,

> there is no discussion of how social activity becomes meaningful to the individual. In discussing language, Vygotsky stresses its social

origins and organising functions but does not consider the influence of the self-generative activities of the individual on linguistic expression. (Fireman and Kose, 1990, p.17)

Vygotsky, living and writing in the climate of dialectical materialism, takes for granted that things are what they are and that 'in reality the child is guided by the concrete, visible likeness' to form associative complexes, or 'pseudo-concepts' which are then modified and tuned by 'verbal intercourse with adults' until the words associated with them 'mean the same to him and to the adult'. This last assertion is the only one I do not agree with. For Vygotsky, 'the same' (in this context) meant the real-world referents of the words. I do not know whether he actually believed in this profession of realism or felt he had to make it because of the absolute Marxist orthodoxy required by the political system under which he had to work. Whatever his reasons were, the statement is unacceptable from the point of view I have been explaining: what a word means is always something an individual has abstracted from his or her experience — it may prove to be compatible with the abstraction another has made, but it can never be shown to be the same. Inevitably, from this perspective, the usual notion of understanding has to be modified.

Interpreting a communication is the process of weaving a conceptual web such that it satisfies the constraints that are indicated by the received linguistic items. Insofar as, given the words heard or read, the receivers succeed in completing a coherent conceptual structure, they will consider that they have understood the piece of language.[4] The linguistic items do not supply the conceptual material, but they delimit what is eligible. In English, for instance, almost every word, taken as an isolated item, has more than one meaning. When it is said or written in a sentence, however, the context of communication usually eliminates all but one of the potential meanings. Instances of irresolvable ambiguity are remarkably rare. In this sense the linguistic and the situational context have a selectional function, much as in the theory of evolution nature or the environment selects viable organisms by eliminating others (what survives, does so, because it has the wherewithal to cope with, and thus to fit into, the environmental constraints).

In communication, the result of an interpretation survives and is taken as the meaning, if it makes sense in the conceptual environment which the interpreter derives from the given words and the situational context in which they are now encountered. The constraints that are inherent in conceptual environments are, of course, far less tangible and definite than degrees of temperature and humidity, speed of locomotion, rate of reproduction, etc., which are the factors that delimit an organism's potential for survival. Nevertheless, opaque though the conceptual conditions may often be, they do determine whether or not a word or a sentence can be fitted meaningfully into the web of an interpretation and whether or not that interpretation can be fitted into the context of the interpreter's general experience. The point to be emphasized is that neither in the realm of evolution nor in linguistic inter-pretation do the constraints specify the actual properties of the items that can or cannot fit into the allowed space. The constraints eliminate what does not fit, but in no way interfere with what does not conflict with them.

There are times when the constraints seem to be remarkably loose. William Lax, a family therapist, recently reported the following episode.

> Client (*spontaneously*): 'Do you remember what you said to me about X the last time?'
>
> W.L. (*waits and says nothing, because he is sure that he never mentioned X during the last session*) . . .
>
> Client: 'It was enormously helpful to me.' (Lax, 1993)

To put it as simply as possible, to 'understand' what someone has said or written implies no less but also no more than to have built up a conceptual structure from an exchange of language, and, in the given context, this struc-ture is deemed to be compatible with what the speaker appears to have had in mind. This compatibility, however, cannot be tested by a direct compar-ison — it manifests itself in no other way than that the speaker subsequently says and does nothing that contravenes the expectations the listener derives from his or her interpretation.

From this perspective, there is an intrinsic, inescapable indeterminacy in linguistic communication. Among proficient speakers of a language, the individual idiosyncrasies of conceptual construction rarely surface as long as the topics of communication are everyday objects and events that have been frequently experienced and talked about by everyone concerned. However, when a conversation turns to predominantly abstract matters, it usually does not take long before conceptual discrepancies become noticeable and generate perturbations in the interaction. At that point the difficulties often become insurmountable if the participants believe that their meanings of the words they have used are fixed, independent entities in an objective world that is the same for all speakers. If, however, the participants adopt a constructivist view and begin by assuming that a speaker's meanings cannot be anything but subjective constructs, a productive accommodation and adaptation can mostly be reached.

Why Communication? Why Language?

If the process of language acquisition is as arduous a business as I have de-scribed, one might ask what is its incentive and its pay-off. There are many ways to approach this question. Given that, as far as we can look back in his-tory, children had practically no choice in the matter: acquiring the language of their group was no less imperative than learning to walk on two legs. However, the question why human or hominid groups developed language is still as unresolved as it was at the end of the last century, when the French Academy banned studies on the origin of language because it was tired of publishing fantasies that no one could ever confirm.

The communication experiments with great apes during the last three decades revived the topic. Having scratched the surface of contemporary primatology during my work with the chimpanzee Lana, I found no reason to believe that the beginnings of linguistic communication made the struggle for survival so much easier for anthropoid apes that it could have become a feature in the process of biological evolution. Rather, I suggested that it began with juvenile play and, from there, developed into an important factor in social evolution (Glasersfeld, 1980, 1992a).

Be this as it may, from the constructivist perspective, we need a psychological model to make plausible the child's effort to acquire language. No doubt it needs to be a multifaceted model or several that function in coordination. I want to propose just one possible facet.

One of our basic assumptions is that the living organism in the struggle to generate and maintain its equilibrium tries to establish regularities in the flow of experience. This manifests itself in the 'circular reactions' that Piaget observed and described as a salient feature of development during the child's first year of life (Piaget, 1937). The term refers to an infant's attempts to repeat an action, because of the perceptual event that followed upon it in a chance experience. If the action again 'produces' the event, this seems gratifying to the infant and he or she will repeat the procedure until something else catches his or her attention.

Other psychologists have called examples of infants' active interference with their experience 'early learning'. One experiment is particularly striking. A pressure switch is placed under a baby's pillow so that it will turn on a light whenever the baby turns its head to the right. Then, if an accidental movement of the baby's head switches on the light, the baby will repeat the movement and the apparently interesting sensory experience until it tires of the effect (Lipsitt, 1966).

This fits nicely with the assumption that action schemes are constructed in order to gain some control over experience. It seems to be a pattern that is applicable in a great variety of circumstances. Another feat of early learning would seem to be based on the repeated experience that crying leads to an improvement of the situation, be it that an unpleasant state is alleviated or boredom ended by the pleasure of being picked up. As life goes on, there are innumerable occasions for the construction of such primitive control schemes. I call them 'primitive', because as soon as some mastery of words is acquired, it opens up a far richer and more powerful field for the generation of predictable reactions among the others in one's environment.

There is a great variety in the ways adults use words. They may be used as 'signals' to induce an action or gain access to an object. There are trivial occasions when they are used gratuitously, without purpose. In the overwhelming majority, words are used instrumentally and serve some goal. No matter how the goals differ, they have one thing in common: they are aimed at the conceptual structures of the receiver. A question is intended to call forth the formulation of a piece of knowledge the other is believed to have. Comments, descriptions, explanations, and lectures aim at modifying the listeners' experiential reality, and sometimes an utterance is designed to call forth memories or emotions. In all cases, the first requirement of the successful use of language is the receiver's effort to construct an interpretation.

Like most of the ideas that make up radical constructivism, this is not new, but it needs repeating. Almost fifty years after her seminal work on the conceptual mechanisms in reading, the many misinterpretations of her theory drove Louise Rosenblatt to state once again her basic principle. Though she was mainly concerned with literature, what she says about the 'object' in reading applies equally to texts in science and philosophy and may serve as a fitting conclusion to my remarks about the general purposiveness of language.

> The 'object' on which the aesthetic reader concentrates is not 'verbal,'
> but experiential; the 'object' is the cognitive and affective structure
> which the reader calls forth and lives through. (Rosenblatt, 1985,
> p.102)

Notes

1 My use of the word 'symbol' is different from both de Saussure's and Piaget's for whom symbols always have an 'iconic' relation to what they symbolize, i.e., they require some likeness between the symbol and its meaning.
2 This primary function of words is called 'denotation', but I avoid this term because it usually implies reference to real-world objects and in my model of semantics words 'refer' to pieces of experience.
3 *Gestalt* psychology investigated this process and came up with several useful hypothetical principles. They have been largely disregarded by the psychological establishment because of the still dominant realist orientation, and for realists there is no problem about perceiving *things* — they are what they are.
4 With the exception of science fiction and freakish tales, understanding usually implies that the conceptual structure can also be fitted into the framework of their experiential world.

Chapter 8

The Cybernetic Connection

The term 'cybernetics' has been used in different ways in popular articles and books and also in the technical literature. It has become a fairly general term that crops up in a variety of contexts. It still has, I suspect, a vague, somewhat mysterious meaning for most readers. This is not surprising, because the cyberneticians themselves have different individual perspectives. Some years ago I was asked by the American Society for Cybernetics to write a description of their field. I could do no better now and am therefore using this 'Declaration' as the first part of this chapter.

The second part is a paper that makes connections between a specific area of cybernetics, Piaget's theory of cognition, and constructivist epistemology.

The last section deals with the particular notion of 'model' that I have used throughout this text. This word, too, is used in many contexts and has a more or less special meaning in each. I have used it frequently in my papers, but did not always announce that I was borrowing it from cybernetics. This has led to misunderstandings. The last part of the chapter should put this right.

Declaration of the American Society for Cybernetics[1]

Cybernetics is a way of thinking, not a collection of facts. Thinking involves concepts: forming them and relating them to each other. Some of the concepts that characterize cybernetics have been about for a long time, implicitly or explicitly. Self-regulation and control, autonomy and communication, for example, are certainly not new in ordinary language, but they did not figure as central terms in any science.

Self-regulation was ingeniously implemented in water clocks and self-feeding oil lamps several hundred years BC. In the scientific study of living organisms, however, the concept was not introduced until the nineteenth century and the work of Claude Bernard. It has a long way to go yet, for in psychology, the dogma of a passive organism that is either wholly determined by its environment, or by its genes, is still frequently accepted without question.

It is much the same with the concept of autonomy. Potentates and politicians have been using it ever since the days of Sparta; but the structural and functional balance that creates organismic autonomy has only recently begun to be studied (e.g., Maturana and Varela, 1980). And there is another side to the concept of autonomy: the need to manage with what is available. That this principle governs the construction of human knowledge, and therefore lies at the root of all epistemology, was first suggested at the beginning of the eighteenth century by Vico and then forcefully argued by Kant (see Chapter 2). The implications of that principle are only today being pursued in some of the sciences.

As for communication, its case is perhaps the most extreme. We are told that the serpent communicated with Adam and Eve shortly after they had been created. Moses communicated with God. And ordinary people have been communicating with one another all along. However, a theory of communication was born a mere forty years ago, when cybernetics began (Wiener, 1948; Shannon, 1948). It was, however, still an observer's theory and said nothing about the requisite history of social interactions from which alone the communicators meaning could spring. Cybernetics arose when the notions of self-regulation, autonomy, and hierarchies of organization and functioning inside organisms were analysed theoretically, that is, logically, mathematically, and conceptually. The results of these analyses have turned out to be applicable in more than one branch of science.

Cybernetics, thus, is metadisciplinary, which is different from interdisciplinary, in that it distils and clarifies notions and conceptual patterns that open new pathways of understanding in a great many areas of experience.

The investigation of self-regulation, autonomy, and hierarchical arrangements led to the crystallization of concepts such as circular causality, feedback, equilibrium, adaptation, control, and, most important perhaps, the concepts of function, system, and model. Most of these terms are popular, some have become fashion words, and they crop up in many contexts. But let there be no mistake about it: the mere use of one or two or even all of them must not be taken as evidence of cybernetical thinking. What constitutes cybernetics is the systematic interrelation of the concepts that have been shaped and associated with these terms in an interdisciplinary analysis which, today, is by no means finished.

Whenever something is characterized by the particular interrelation of several elements, it is difficult to describe. Language is necessarily linear. Interrelated complexes are not. Each of the scientists who have initiated, shaped, and nourished this new way of thinking would describe cybernetics differently, and each has defined it on a personal level. Yet they are all profoundly aware that their efforts, their methods, and their goals have led them beyond the bounds of the traditional disciplines in which they started, and that, nevertheless, there is far more overlap than individual divergence in their thinking. It was Norbert Wiener (1948), a mathematician, engineer, and social philosopher, who adopted the word 'cybernetics'. Ampère, long before, had suggested it for the science of government, because it derives from the Greek word for 'steersman'. Wiener, instead, defined cybernetics as the science of 'control and communication in the animal and the machine'. For Warren McCulloch, a neuroanatomist, logician, and philosopher, cybernetics was experimental epistemology concerned with the generation of knowledge through communication within an observer and between observer and environment. Stafford Beer, industrial analyst and management consultant, defined cybernetics as the science of effective organization. The anthropologist Gregory Bateson stressed that whereas science had previously dealt with matter and energy, the new science of cybernetics focuses on form and patterns. For the educational theorist Gordon Pask, cybernetics is the art of manipulating defensible metaphors, showing how they may be constructed and what can be inferred as a result of their construction. And we may add that Jean Piaget, late in his life, came to see cybernetics as the endeavour to model the processes of cognitive adaptation in the human mind.

Two major orientations have lived side by side in cybernetics from the beginning. One is concerned with the conception and design of technological developments based on mechanisms of self-regulation by means of feedback and circular causality. Among its results are industrial robots, automatic pilots, all sorts of other automata, and of course computers. Computers, in turn, have led to the development of functional models of more or less intelligent processes. This has created the field of artificial intelligence, a field that today comprises not only systematic studies in problem solving, theorem proving, number theory, and other areas of logic and mathematics, but also sophisticated models of inferential processes, semantic networks, and skills such as chess playing and the interpretation of natural language.

Other results of this essentially practical orientation have been attained in management theory and political science. In both these disciplines cybernetics has elaborated principles that clarify and systematize the relations between the controller and the controlled, the government and the governed, so that today there is a basis of well-defined theories of regulation and control (Ashby, 1952; Conant, 1981; Powers, 1973).

The other orientation has focused on the general human question concerning knowledge and, placing it within the conceptual framework of self-organization, has produced, on the one hand, a comprehensive biology of cognition in living organisms (Maturana and Varela, 1980) and, on the other, a theory of knowledge construction that successfully avoids both the absurdities of solipsism and the fatal contradictions of realism (von Foerster, 1973; McCulloch, 1970; Glasersfeld, 1976b).

Any attempt to know how we come to know is obviously self-referential. In traditional philosophy and logic, crude manifestations of self-reference have always been considered to be an anomaly, a paradox, or simply a breach of good form. Yet, in some areas, processes in which a state reproduces itself have been domesticated and formally encapsulated; and they have proven extremely useful (e.g., eigenvalues in recursive function theory, certain topological models derived from Poincaré, condensation rules in logic, and certain options in programming languages for computers, especially for application to non-numeric computations such as in knowledge engineering and expert systems). The formal management of self-reference was dramatically advanced by Spencer Brown's calculus of indications (1973), in which the act of distinguishing is seen as the foundation of all kinds of relationships that can be described, including the relationships of formal logic. Recent studies, building on that foundation and extending into various branches of mathematics, have thrown a new light on the phenomenon of self-reference (Varela, 1975; Goguen, 1975; Kauffman, 1987).

The epistemological implications of self-reference have an even wider range of influence in the cybernetical approach to the philosophy of science. Here there is a direct conflict with a tenet of the traditional scientific dogma, namely the belief that scientific descriptions and explanations should, and indeed can, approximate the structure of an objective reality, a reality supposed to exist as such, irrespective of any observer. Cybernetics, given its fundamental notions of self-regulation, autonomy, and the informationally closed character of cognitive organisms, encourages an alternative view. According to this view, reality is an interactive conception because observer and observed are a mutually dependent couple. Objectivity in the traditional sense, as Heinz von Foerster has remarked, is the cognitive version of the physiological blind spot: we do not see what we do not see. Objectivity is a subject's delusion that observing can be done without him. Invoking objectivity is abrogating responsibility — hence its popularity.

Observer–observed problems have surfaced in the social sciences with the emergence of the notion of understanding. In anthropology, for example, it has been realized that it is a sterile undertaking to analyse the structure of a foreign culture, unless a serious effort is made to understand that culture in terms of the conceptual structures that have created it. Similarly, in the study of foreign or historical literature, the hermeneutic approach has been gaining ground. Here, again, the aim is to reconstruct meaning in terms of the concepts and the conceptual climate at the time and the place of the author. The emerging attitude in these disciplines, though traditionalists may be reluctant to call it scientific, is in accord with cybernetical thinking.

The most powerful and encouraging corroboration of the cybernetician's disengagement from the dogma of objectivity, however, comes from the hardest of the sciences. In physics, the problem of the observer reared its head early in this century. The theories of relativity and quantum mechanics almost immediately raised the question of whether they actually pertained to an objective reality or, rather, to a world determined by observation. For some time the question was not answered definitively. Einstein was hoping that the realist interpretation would eventually lead to a homogeneous view of the universe. Heisenberg and Bohr tended the other way. The most recent in the long series of particle experiments have lessened the chances of realism. Realism in this context was the belief that particles, before anyone observes them, are what they are observed to be. Physics, of course, is not at an end.

New models may be conceived, and the notion of an objective, observer-independent reality may once more come to the fore. But at present, the physicist's theories and experiments confirm the cybernetician's view that knowledge must not be taken to be a picture of objective reality but rather as a particular way of organizing experience.

In the few decades since its inception, cybernetics has revolutionized large areas of engineering and technology. Self-regulation has moved from the refrigerator into the cars we drive and the planes we fly in. It has made possible the launching of satellites and 'explorers' of our solar system. It has also saddled us with target-seeking missiles, and it has brought about the computer age with its glories and its dangers.

For many of us, however, this explosion of gadgetry is not the most significant feature. The wheel, the harnessing of electricity, the invention of antiseptics and the printing press have all had somewhat similar effects on the mechanics of living. Cybernetics has a far more fundamental potential. Its concepts of self-regulation, autonomy, and interactive adaptation provide, for the first time in the history of western civilisation, a rigorous theoretical basis for the achievement of dynamic equilibrium between human individuals, groups, and societies. Looking at the world today, it would be difficult not to conclude that a way of thinking which, rather than foster competition and conflict, deliberately aims at adaptation and collaboration may be the only way to maintain human life on this planet.

Feedback, Induction, and Epistemology[2]

One of the most successful notions in control theory has been the principle of negative feedback. As Otto Mayr shows in his delightful book *The Origins of Feedback Control* (1970), practical implementations of the principle go back to the third century BC, explicitly documented in the case of oil lamps that regulate the flow of oil according to the amount they burn. Today we have thermostats, automatic pilots, and guided missiles. Though these devices differ in structure and material, they have one thing in common: within certain limits they are able to carry out activities that formerly required a human agent's attention, discrimination, and judgment. All control mechanisms were designed to free someone's hands or mind for a more important task or, perhaps, just for a more entertaining activity. From the very beginning, their purpose was to maintain or create some state which the designer or user deemed desirable in his or her experiential world — to keep a lamp burning after the slaves were sent to bed, to keep the room at an even temperature regardless of the weather, and so on. All this is taken for granted today, and it is one reason why we are prone to overlook some basic aspects of the phenomenon. As Powers (1978) demonstrated, the embeddedness in the user's goal structures has led to misinterpretations of how feedback mechanisms actually function in living organisms.

The feature I want to focus on here is that 'control systems . . . control input, not output'.

> Natural systems cannot be organised around objective effects of their behaviour in the external world; their behaviour is not a show put on for the benefit of an observer or to fulfil an observer's purpose. A natural system can be organised only around the effects that its actions (or independent events) have on its inputs . . . (Powers, 1978, p.418)

From the constructivist perspective, 'input' is of course not what an external agent or world puts in, but what the system experiences. This can be shown in a very simple diagram I have adapted from Powers (1973, p.61).

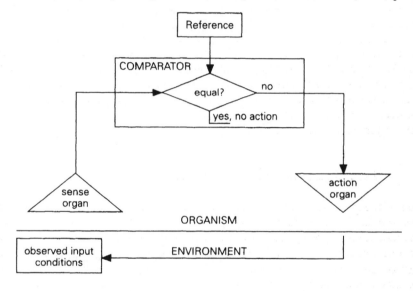

Figure 8.1:   The Feedback Loop (after Powers)

A control system acts when there is a discrepancy (negative feedback) between what it senses (sensory signal) and what it is supposed to sense or would like to sense (reference). Only an observer is in a position to say that an action of the organism changes conditions in its environment and therefore what it senses. The organism itself merely reacts to a discrepancy between the reference value and what it senses. If it happens to be in an environment where its actions have no effect on what it senses, the discrepancy may get larger and larger and become fatal. The relations that matter to the organism are those between the activities in its repertoire and the changes they provoke in its sensory perturbations.

A mechanical feedback device that replaces us in a given task is a crystallized piece of our own experiential learning. It is the materialization of an if-then rule that was inductively derived from experience by the designer.

A Learning Mechanism

Let us, for a moment entertain the fanciful assumption that the thermostat of an air-conditioning system were miraculously imbued with awareness and some cognitive functions so that it could think about and organize its experiential world. It would be a very simple world. The only perceptual discriminations the thermostat could make would be between signals from its thermometer that fall short of the reference temperature, signals that match it, and signals that are in excess of it. There could be no other perceptual data. On the proprioceptive side, i.e., the system's kinaesthetic feedback generated by its own acting, the activity of heating could be discriminated from the activity of cooling. In other words, all the thermostat could come to know in its experiential world would be that it feels too hot or too cold and whether it is at the moment exercising its heating or its cooling activity. The connections between the two kinds of perceptual perturbation and the activities are fixed. These connections are similar in that respect to those implied by reflexes or fixed action patterns in living organisms. Neither in the thermostatic control device nor in the organismic reflex did those connections require learning on the part of the individual system that manifests them. They are wired in, by the designing engineer in the case of the device, and, in the case of the organism by evolution, through the processes of variation and selection.

In a more complex system, however, the connections may be the result of learning. Kenneth Craik, a precursor of cybernetic thinking in the early 1940s, suggested how an elementary form of learning could be mechanized (Craik, 1966). It requires two things: on the one hand, something like a memory, a place where sequences of signals could be recorded to be read at some later point in the experiential flow; on the other, the ability to compare past signals to present ones or to a goal-signal that constitutes a reference value. Once that dual capability is there, the preconditions of inductive learning are satisfied. On this initial level, induction is as simple as it was described 250 years ago by David Hume (1742). All that is needed is the disposition or rule that leads the system to repeat actions that were recorded as successful in its past experience. That is to say, in each occurrence of a perturbation, the system will select the activity that reduced or eliminated that specific kind of perturbation in the past. Implicitly or explicitly, there must be the belief that connections that turned out to be successful, will be successful also in the future. For, as Hume said:

> if there be any Suspicion, that the Course of Nature may change, and that the past may be no Rule for the future, all Experience becomes useless, and can give rise to no Inferences or Conclusions. (Hume, 1742, Essay II, Part 2)

No matter how sophisticated the cognitive functions we hypothetically attribute to an imaginary learning thermostat, it could never do more than establish regularities concerning specific connections between its activities and the subsequently experienced changes of sensory signals. It could not discover that by activating its heating machinery it changes the temperature in the environment. All it could learn would be that its heating activity reduces the sensation of cold and the cooling activity the sensation of heat. It could learn to control its perceptions. That there is an external connection could be specified only by an observer, because from an observer's point of view both the organism and its environment are segments of actual experience. From the organism's perspective, whatever connections are made and whatever regularities are found, are always connections and regularities of its own internal signals.

Cognitive Development

The theory of cognitive development that was proposed and elaborated by Piaget has deep biological roots and builds on presuppositions that are intended to apply to all forms of life. Perhaps the most important among these presuppositions is the assumption that what differentiates living organisms from the rest of the universe is their concern with an inner milieu and their relative ability actively to maintain internal states in equilibrium in spite of external perturbations. All activity — and thus also cognitive activity — is considered adaptive in the specific sense that it serves the purpose of self-regulation (e.g., Piaget, 1967a).

The biological organism does not begin life as a *tabula rasa*. We need not claim that it starts out with god-given Platonic ideas or with genetically transmitted knowledge of an outside world. It is sufficient to assume that the organism has a tendency to act in the face of perturbation. Piaget's key to development, i.e., the increase of internal organization, is the concept of 'scheme'. Regardless of whether a scheme is implemented in a reflex or a sophisticated arrangement of cognitive structures, it consists of three parts. First, as I laid out in Chapter 3, there is a pattern of sensory signals which, from an observer's point of view, may be considered the effect of an external stimulus; second, there is an activity, triggered by the particular pattern of sensory signals and which an observer may consider a response; third, subsequent to the activity, the organism experiences some change which, sooner or later, is registered as the consequence of the activity. The consequence is in fact the reason why particular activities are linked to particular perturbations.

The Inductive Basis of Instrumental Learning

On the evolutionary level, natural selection tends to eliminate individuals that have non-adaptive reactions to perturbations from the environment, whereas those that happen to have adaptive reactions survive. Phylogenesis, thus produces results which, considered retrospectively, look as though they were the result of induction: what survives are only those mutants that happen to weather the perturbations of the environment.

On the ontogenetic level, the pattern is similar. The 'Law of Effect', 'Other things being equal, connections grow stronger if they issue in satisfying states of affairs' (Thorndike, 1931), is essentially equivalent to the paradigm:

> The living system, due to its circular organisation, is an inductive system and functions always in a predictive manner: what occurred once will occur again. Its organisation (both genetic and otherwise) is conservative and repeats only that which works. (Maturana, 1970b, p.39).

For Maturana, speaking as a biologist, the expression 'it works' means that, what the system does, successfully eliminates a life-threatening perturbation.

However, the same inductive principle is inherent also in Piaget's concept of 'scheme', but there it is a principle of cognition. Schemes serve not only biological survival but also organisms' cognitive goals whose non-attainment is not fatal. They are part of an instrumentalist theory of learning and incorporate the processes of assimilation and accommodation.

In order to be activated, a scheme requires the perception of a particular pattern of sensory signals. In actual experience, however, no two situations are quite the same. The sensory pattern that triggers a particular scheme must, therefore, be isolated by the organism in a perceptual field that usually provides vastly more signals than those needed for the particular pattern. At other times the perceptual field does not provide all the necessary signals. In other words, differences must be disregarded, and this disregarding of differences, so that the pattern can be obtained in spite of them, is called assimilation.

The acting system or organism, does not notice specific differences because it is looking for the signals required to complete a pattern that might trigger a scheme. In contrast, an observer who does register extraneous signals could say that the organism is assimilating (see Chapter 3).

Sophisticated cognitive organisms, however, have the capability to disregard such differences deliberately. For them, assimilation becomes a crucial instrument in the construction of regularities and rules, as well as for the practical extension of their schemes. To give an example, if Mr Smith urgently needs a screwdriver to repair the light switch in the kitchen, but does not want to go and look for one in his basement, he may 'assimilate' a butter knife to the role of tool in the context of that particular repair scheme. This is not quite like an infant's assimilation, because Mr Smith remains aware of the fact that the butter knife is perceptually and functionally different from a screwdriver.

Whenever a scheme is activated and the triggered activity does not yield the expected result (e.g., Mr Smith's butter knife bends and does not turn the screw), the discrepancy from the accustomed sequence of events creates a perturbation in the system. As this perturbation springs from the mismatch of an actually sensed situation and an expected one that served as reference, it is equivalent to negative feedback in a cybernetical control loop. It is a novel kind of perturbation. It is not associated with a specific sensory pattern, nor an activity that might eliminate it. However, because it arises as the result of an enacted scheme, it may direct the agent's attention to the sensory material that was present when the scheme was activated (see Piaget, 1974a, p.264) and this may then lead to an accommodation of the scheme or the formation of a new one (see Chapter 3).

As in the case of assimilation, such an accommodation may take place without the agent's awareness, or it may be deliberate. Every time we sit down on an unfamiliar chair, the physical movements that constitute the motor part of our sitting-down scheme may have to be slightly adjusted to fit the particular circumstances, but we usually remain quite unaware of that accommodation. When, on the other hand, we drive a new car, we also have to make certain adjustments: we deliberately accommodate our motor acts and sometimes even construct (by trial and error) novel subschemes to fit into, or partially replace, the ones we had.

Negative Feedback as 'Information'

Such sensorimotor schemes constitute the lowest but nevertheless essential level of cognitive development; and the concepts of scheme, assimilation, and accommodation are no less applicable to the higher levels of cognition.

From the system's point of view, the conception of the scheme with its inherent processes of assimilation and accommodation and the conception of the learning feedback mechanism are analogous and wholly compatible. In both cases, all vital knowledge is constituted by rules that indicate which particular actions are successful in eliminating particular perturbations. No knowledge of an independent external reality is gained, nor is any such knowledge needed.

Analogously to a learning cybernetic system, a living organism must be able to experiment and to construct, by inductive learning from experimental outcomes, a repertoire of schemes that enable it to maintain its sensory perceptions within an acceptable range of the reference values.

The situation is similar to that of living organisms in the theory of evolution. Only the viable biological structures survive, because natural selection does away with organisms that cannot in some way avoid or compensate for the environmental perturbations. Avoidance and compensation are the means to maintain an equilibrium.

Gregory Bateson was the first to make the connection between cybernetics and the theory of evolution:

> Cybernetic explanation is always negative. We consider what alternative possibilities could conceivably have occurred and then ask why many of the alternatives were not followed, so that the particular event was one of those few which could, in fact, occur. The classical example of this type of explanation is the theory of evolution under natural selection. According to this theory, those organisms which were not both physiologically and environmentally viable could not possibly have lived to reproduce. Therefore, evolution always followed the pathways of viability. (Bateson, 1972b, p.399)

On the cognitive level, as a rule, the perturbations are not immediately fatal. Phylogeny proceeds by pruning; ontogeny provides opportunities for learning. In both domains, organisms may meet reality only in their failures. As Warren McCulloch said: 'To have proved a hypothesis false is indeed the peak of knowledge' (McCulloch, 1970, p.154). This is equivalent to negative feedback: things are not what we thought they were.

From the perspective of traditional epistemology, McCulloch's statement is a declaration of the discipline's bankruptcy. Ever since the pre-Socratics, knowledge was supposed to correspond to a real world. If it did, it was true, if it did not, it was worthless. The notion of viability within constraints, is incompatible with the conventional one of truth and correspondence.

If one takes seriously the proposition that cognitive organisms do not make contact with an ontological reality *except* when their schemes to eliminate perturbations break down, one can come to a more positive albeit less metaphysical view of knowledge. In the domain of schemes that involve action, their value has always been assessed on the basis of whether or not they achieve what they are expected to achieve. In other words, it is a question of know-how, and know-how has functional value. As with all functional values, further criteria, such as amount of effort and cost, speed, or the aspect of elegance, and other features can be added.

Functional values, however, are not the only ones. With the construction of schemes the first step is made into a virtually infinite hierarchy of levels of reflection and abstraction. Although cognitive structures and schemes never lose all connection to the functional level of action at the bottom of the ladder, their assessment comes to involve criteria of homogeneity, compatibility, and consistency, as one moves up the rungs of abstraction. The crucial aspect of our theory of knowing is that the idea of correspondence with reality is replaced by the idea of fit. Knowledge is good knowledge if it fits within the constraints of experiential reality and does not collide with them. This fit must be attained not only insofar as a cognitive structure, a scheme, a theory, have to remain viable in the face of new experience or experiments, but also in that they prove compatible with the other schemes and theories one is using.

This aspect of cybernetical epistemology may seem similar to Popper's (1968) principle of 'conjectures and refutations', but there are important differences: we put the stress on the viability of the conjectures rather than on their refutation; and we do not claim that the pursuit of viability is a progression towards truth.

Neither in the realm of evolution, nor in that of the interpretation of experience, do the constraints one encounters determine actual properties of the items that do or could fit into the allowed space. The constraints merely eliminate what does not fit.

Norbert Wiener's definition of cybernetics hinges on the concepts of control and communication. While he viewed control mechanisms mainly from the perspective of engineers who use feedback devices as proxies for themselves, he did not stress the epistemological implications that arise if one considers these devices as independent, self-regulating systems. There is no contradiction between an engineer's use of a feedback control gadget and the learning organism I have outlined. The engineer's gadget is the outcome and an extension of the engineer's experiential world — the organism, in contrast, has its very own subjective experience.

The Nature of Hypothetical Models

There is one more aspect of cybernetics that relates it to theories of cognition, namely the endeavour to construct actual or conceptual models that simulate the functional properties of a black box. The formalistic branch of the discipline aims at the development of mathematical models, i.e., networks of functions that mathematically account for and predict observable output from observable input. The more concrete, heuristic branch of the discipline aims at the development of conceptual or physical models that are operationally equivalent to the unobservable mechanisms inside a black box. In both these branches of cybernetics one works towards a fit and not towards an iconic replication. Hence, a model is a good model whenever the results of its functioning show no discrepancy relative to the functioning of the black box. That relation, I claim, is analogous to the relation between our knowledge and our experience. Given that there is nothing but a hypothetical connection between our experience and what philosophers call ontological reality, that reality has for us the status of a black box.

One of the characteristics of cybernetics is that it produces 'explanations' which, as Bateson said, do not specify why certain things happen, but rather why other possible things did not happen. It specifies constraints. Cyberneticians, however, often take a further step: they think up a functional model that would produce effects that are similar to those of the observed phenomenon. This is a useful way of trying to gain some conceptual or practical control over a mechanism that is inaccessible to observation and, therefore, what cyberneticians call a 'black box' (see Wiener, 1965, p.XI). The feedback loop is such a model that has been highly successful. It has made possible the mechanization of all sorts of things that formerly, only a human being could do. This very success, however, has often led to the hasty conclusion that the feedback loop is a description of how humans or animals actually function. (For research in artificial intelligence it seems particularly difficult to avoid such derailments.)

Successful models fit and function within the constraints set by the given situation, but there is never any reason to believe that they embody the only mechanism or conceptual network that could do this. Hence there are three points that have to be remembered with regard to models of the cybernetic kind — and they are especially pertinent to the constructivist view.

First, we must remain constantly aware of our basic assumption that concepts and conceptual structures are necessarily hypothetical items. They are doubly hypothetical whenever they are attributed to others. At best we can know them to the extent that the owner or user tells us about them or, alternatively, acts in a way that leads us to infer them. Both these ways of access, however, are subject to a general restriction which, although it is traditionally disregarded by realists of every denomination, must be taken very seriously by constructivists. In its simplest form, the restriction amounts to this: whenever we interpret what others say, or the way they act, we interpret what we hear or see in terms of elements that are part of our own experience. We cannot have another's experience.

To use a drastic example, a congenitally blind person's interpretation of his or her sighted friends is necessarily composed of elements within the blind person's domain of experience. This interpretation may contain correlations, regularities, and probabilities that are different from those the person has previously constructed, but it cannot possibly contain elements that derive from visual experience.

Second, there is the purpose of hypothetical models. If we are not satisfied with mere descriptions of observable behaviour but want to formulate hypotheses as to how the observed behaviours come about, the simplest procedure would be to open up the behaving organism in order to see what goes on inside. Living organisms, however, have the awkward peculiarity that their more interesting functions cease when we cut them open. From a cybernetician's point of view, therefore, living organisms (children and students in particular) are black boxes. Their internal functions are not accessible to observation. Yet, much like cyberneticians, psychologists and educational researchers want to go further: they want to see if they can set up hypothetical operations that would yield the same results as those manifested as behaviour by the observed organism. Although these models are hypothetical and must never be said to depict or replicate what actually goes on, they may be extremely useful. After all, it is better to know at least one way in which a given behaviour (or reply) *could* be produced, than to have no idea at all.

This way of proceeding is, in fact, not very different from that of the modern physicists who, in order to construct theories and make predictions about observable events, postulate hypothetical entities with hypothetical properties that lie beyond the range of direct observation.

Third, there is the problem of development. When some phenomenon is to be explained developmentally, differences must be found between what the organism is doing now, and what was observed before or will be observed at a later time. If such differences are found, they have to be interpreted. In the context of development, the differences are always interpreted in view of what, from the observer's point of view, is being developed. That is to say, there is a guiding idea of an end-state or target product. If there were no such idea, it would simply be a study of change. Moreover, when we speak of the development of children (or students), we have the ultimate goal of specifying a plausible succession of changes that should characterize a generalizable progression from an original (primitive) way of acting to the accepted adult way of acting or responding to certain experiential situations.

I originally made these three points as explicit as I could, as an admonition to myself. Here they may serve to forestall any realist interpretation of what I am presenting. They are pertinent to the whole of radical constructivism because it, too, is a model that does not purport to be the description of any reality. If it turns out to be compatible with such observations as have been or will be made, it will be a viable model that can be used to make predictions and as guideline in a variety of areas including the development of didactic methods.

Notes

1   The following people have contributed ideas, formulations, and critical suggestions to this document: Stuart Umpleby, Paul Trachtman, Ranulph Glanville, Francisco Varela, Joseph Goguen, Bill Reckmeyer, Heinz von Foerster, Valentin Turchin, and my wife Charlotte. I alone, however, should be held responsible for the shortcomings of this survey.
2   Revised and expanded from Glasersfeld (1981b) (Courtesy Pergamon Press).

Chapter 9

Units, Plurality and Number[1]

For some fifty years Piaget was saying that the process of perception does not seem feasible unless we assume that the perceiver has some prior structure to which he can assimilate his sensory experience. Though there are empirical findings that corroborate this hypothesis, it draws its strength from the epistemological foundation on which Piaget has built his entire theory of cognition. The notion that what we come to know is to a large extent selected and shaped by what we already know, has cropped up independently in the philosophy of science.

A century ago, most scientists and ordinary people believed that what they called 'data' was there to be found by anyone who looked closely enough. This belief has been shaken. Today, a new generation of scientists is more inclined to think that the finding of data presupposes a specific theoretical structure to direct and inform search and observation. Hanson (1958, p.19) said it very simply: 'Observation of x is shaped by prior knowledge of x' (see also Bridgman, 1961; Kuhn, 1962; Feyerabend, 1975). This view was couched in the phrase 'All data is theory-laden'.

Nonetheless, there is still a widespread belief that good data has to be objective and, therefore, independent of any observer's perceptual habits, theories, and beliefs. How, otherwise, could data serve as the material from which a true representation of the environment can be produced? Throughout this text, I have argued that this belief is not a useful one because it leads to a paradox in epistemology and hence to an unsatisfactory model of cognition. I suggested that the constructivist view provides a more promising approach by positing that all knowledge is constructed from subjective experience. This might appear to be quite incompatible with the experiential fact that mathematics produces a host of results that are eminently 'objective' in the sense that no individual subject can question them. Clearly, this is a problem that has to be resolved before the constructivist model can claim to be viable.

An Elusive Definition

Consequently, I shall outline a constructivist method of conceptual analysis, and apply it to the three concepts that are basic to the development of arithmetic and mathematics: unit, plurality, and number. What I am going to present is a hypothetical model and a sequence of conceptual steps that *could* yield the items in question. There is, of course, much more to mathematics than these three concepts, but it would take a mathematician to deal with the many-layered tower of abstractions that has been built on the basis of the elementary arithmetical concepts. My purpose is the very limited one of showing that the three fundamental concepts can be seen as constructs rather than gifts from God or some other source beyond human experience.

I start from the assumption that concepts must somehow be conceived by humans.

Thus much is true, that of natural forms, such as we understand them, quantity is the most abstracted and separable from matter.
Francis Bacon (1623)

Abstracted entities are the result of an abstracting activity carried out by a cognizing subject. Even if one wants to believe that these entities exist independently in some outer space, specific ways of operating would be needed in order to apprehend and *know* them. The constructivist model should show that the three concepts of unit, plurality, and number, which Bacon must have included among his 'natural forms', can be built up *without* taking for granted that they exist ready-made in an objective reality. This does not mean that their construction does not involve perceptual processes. It merely means that the procedure must be constructive rather than passive.

The philosopher Thomas Tymoczko recently suggested that 'mathematics is much more like geography than it is like physics' (1994, p.334). I find this a congenial comparison because I can easily substitute conceptual semantics for the geographer's topography.

Hence, I shall try to do two things: provide an analysis of how the concepts may be structured and suggest that their construction starts from perceptual elements and is achieved by a succession of reflective abstractions. I emphasize that the result of this effort could not be anything but a hypothetical model. It does not purport to be the description of any reality. At best the model may turn out to be compatible with such observations as have been or will be made. If that should be the case, the model would be a useful one, because it could perhaps be used as guideline in the development of didactic methods.

In order to investigate how children form the basic concepts on which arithmetic can be built, it is indispensable to have a fairly explicit model of what these concepts might be in the adult. Mathematics textbooks are not very illuminating in that regard and philosophers of mathematics rarely stoop to say anything about the conceptual raw material of their constructions. The Italian mathematician Giuseppe Peano was an exception. He did not reveal the raw material, but he gave at least a reason why he considered it unnecessary to do so.

The first numbers that present themselves, and with which all others are formed, are integers and positive. The first question is: can we define unity, number, the sum of two numbers? The usual definition of number, which is Euclid's 'number is the aggregate of several units', may serve as clarification but is not satisfactory as definition. In fact, a child of few years uses the words 'one', 'two', 'three', etc.; later it uses the word 'number'; only much later the word 'aggregate' appears in its vocabulary . . . Hence, from a practical point of view, the question seems to me resolved, that is, in the course of instruction it would not be advisable to give any definition of number, since that idea is perfectly clear to the students, and any definition would only have the effect of confusing the idea. (Peano, 1891a, pp.90–1)

He then discusses the theoretical aspects and concludes that number cannot be defined (ibid., p.91).

At the beginning of his essay on the principles of mathematical logic, he put his finger on one of the problems. Signs such as 1, 2, 3/4, $\sqrt{2}$, he explained, are actually proper names, whereas 'number', 'polygon', 'equilateral', etc., refer to classes and are, therefore, common nouns (Peano, 1891b, p.2). This statement is interesting because it clearly brings out the difficulty: individual items have to be characterized by individual characteristics, classes by common ones.

The question, then, is: what are the individual characteristics in the case of 'one', 'two', 'three', and what are the common ones in the case of 'number'?

The kind of question 'What is a . . . ?' can usually be answered in more than one way. In the case of 'number', one answer might be: 'Well, one, two, fifteen, thirty-eight, are numbers. This would be equivalent to answering 'Pippins, Winesap, Golden Delicious', to the question: 'What is an apple?' It would not be much help to a child who has not much experience with apples.

Instead of the verbal reply, one could go to a well-stocked pantry, come back with specimens of Pippin, Winesap, and Golden Delicious, and say: 'All these are apples!' This would be part of what philosophers call an 'extensional' definition. If the child then asked why the things shown are called apples, one could point out that they are relatively round and smooth objects of a certain size and weight, have skin, flesh, and core, and a smell and a taste that one can learn to recognize. In other words, using part of an extensional definition could, by and large, be helpful in specifying some of the characteristics that go to make the concept, and thus produce an 'intensional' definition.

When the question concerns number, there are immediate difficulties. Assuming the child asks 'Why are one, two, fifteen, numbers?', we might start to put one glass, two spoons, and fifteen toothpicks on the table, but we will soon realize that this is unlikely to work. We may then have an inspiration: we push everything aside and arrange toothpicks (or, indeed, multi-base blocks) in lots of one, two, and fifteen. Now, we feel, it should be obvious that we are *showing* numbers. But by then the child would probably want to play another game. This is fortunate, because if a child pursued the question further, we would be stumped by the fact that there is no way of explaining what characteristics one has to abstract in order to form the concept of number. We would have to suggest counting the toothpicks in each pile — but this begs the question because one cannot count without using the words 'one', 'two', 'three', and so on.

Peano and the others involved in the effort to formalize a logical foundation of the number system and mathematics, were intent upon defining properties and relationships *within* the system. They took for granted that we have concepts of unit, plurality, and number. What has to be done to generate these concepts experientially did not seem a problem to them. How else could one explain that a shrewd thinker would come up with the 'definition':

A number is anything which is the number of some class. (Russell, 1956, p.534)

This is circular in the vicious sense, because in order to understand the definition one would have to know the term that is being defined.

Things and Units

What, then, is a number? Maybe Euclid's clarification is helpful, after all. It became clear in the example of the toothpicks, that it is not a characteristic of the individual objects that matters. Perhaps it is their arrangement in lots, that is, their grouping. But this, is not satisfactory either. It merely raises the further question of how we form the concept of group. Besides if four toothpicks are placed one in each corner of the room, they could still be considered as *four*, Where, then, is the necessary aggregation? The answer to that question is both old and frequently disregarded. The earliest statement of it I have found, is also the most elegant and the most convincing. It comes from Juan Caramuel, the seventeenth-century bishop of Vigevano, whom I have cited before (Chapter 5). He told a most revealing anecdote:

There was a man who talked in his sleep. When the clock struck the fourth hour, he said: 'One, one, one, one. That clock must be mad, it has struck one four times.' The man clearly, had counted four times one stroke, not four strokes. He had in mind not a four, but a one taken four times, Which goes to show that to count and to consider several things contemporaneously are different activities. If I had four clocks in my library, and all four were to strike one at the same time, I should not say that they stuck four, but that they struck one four times. This difference is not inherent in the things, independent of the operations of the mind. On the contrary, it depends on the mind of him who counts. The intellect, therefore, does not find numbers but makes them; it considers different things, each distinct in itself, and intentionally unites them in thought. (Caramuel, 1670, pp.xliii–xliv)

For Caramuel it was a matter of 'intentionally uniting [different things] in thought'. Berkeley, some thirty years later, made a note to himself: 'Number not without the mind in anything, because 'tis the mind by considering things as one that makes complex ideas of them' (1706–8, par.106). This, however, still leaves the question where the 'things' come from or, rather, how the mind distinguishes things in such a way that it can unite them to form complex ideas.

McLellan and Dewey had every intention to specify the necessary operations.

In the simple recognition, for example, of three things as three the following intellectual operations are involved: *The recognition of the three objects as forming one connected whole or group* — that is, there must be a recognition of the three things as individuals, and of the one, the unity, the whole, made up of the three things. (McLellan and Dewey, 1908; p.24)

They here used the word 'recognition' — which would imply that oneness or threeness is some kind of perceivable property that belongs to the things — but earlier in their text they pointed out that 'Number is a rational process, not a sense fact' (ibid., p.23); and later, they explain 'that number arises from certain rational processes in construing, defining and relating the material of sense perception' (loc.cit., p.35). Thus, it requires an active mind that takes distinct things and unites them by means of a particular operation. It is these operations of defining and relating that need to be analysed if we are to have an operational model of the number concept.

Clearly, separating and uniting are the crucial activities. There must be an operation that creates discrete unitary items, and there must be an operation that takes several such individual items and unites them so that they can again be seen as a unit. Hence, the first question is: how do we come to have a unit, a unitary item?

Conception Rather than Perception

The physicist Percy Bridgman formulated this question when he asked: What is the thing that we count? He answered it as a constructivist would:

It is obviously not like the objects of common sense experience — the thing that we count was not there before we counted it, but we create it as we go along, It is the acts of creation that we count. (Bridgman, 1961, p.103)

Units, then, are the result of an operation carried out by a perceiving subject, not a property inherent in objects. This may sound absurd, because the adult's habitual and largely automated way of perceiving creates the impression that the unity of an item is *given*, because the item is distinguishable from the experiential background. In fact, it is very likely that infants first derive some notion of thinghood[2] from items that are easy to isolate in the visual field. But distinguishing and isolating are activities that have to be carried out by an active subject, and the results depend on the subject's own criteria of distinction.

As I mentioned (Chapter 1), Husserl explicitly said that the concept of unit is an abstraction from sensorimotor objects. A remarkable confirmation of this idea comes from Albert Einstein:

> I believe that the first step in the setting of a 'real external would' is the formation of the concept of bodily objects and of bodily objects of various kinds. Out of the multitude of our sense experiences we take, mentally and arbitrarily, certain repeatedly occurring complexes of sense impressions (partly in conjunction with sense impressions which are interpreted as signs for sense experiences of others), and we correlate to them a concept — the concept of the bodily object. Considered logically this concept is not identical with the totality of sense impressions referred to; but it is a free creation of the human (or animal) mind. (Einstein, 1954, p.291)

Sensory signals are necessary for the development of the concept of unitary item or thing, in that they provide occasions for the required 'empirical' abstraction. The operations that create the unit, however, are not given in the sensory material but have to be carried out be an active subject (see Humboldt quotation in Chapter 5). A simple example of visual experience can demonstrate this. Looking at Figure 1, you can see the wave line as one continuous unitary item; but you can see it also as three crests or two troughs; and then you can see it as a multitude of discrete unitary dots.

Figure 9.1: Sensory Segmentation

The sensory signals remain the same throughout, yet they can be organized into different kinds of units. The wave line, one might object, nevertheless provides *some* sensory basis for each of the organizations and therefore determines what organizations are possible. But this is an illusion. A straight line, as in Figure 2, can be seen as one unitary piece. Yet, in spite of its perfect sensory homogeneity, it can also be seen as two halves, three thirds, or four quarters.

Figure 9.2: Conceptual Segmentation

If you work a little harder, you can also see that there are roughly five inches in the line or, with more practice in the metric system, about eleven centimetres. With the exception of the unitary segment of line, none of the units is determined by sensory signals. The partitions are a free creation of the perceiver who constructs them by means of a specific unitizing operation.

Before presenting my hypothetical model of the unitizing procedure, I want to point out another kind of operation that manifested its result in the different ways of seeing the wave line. Before you could say that there were *two* troughs or *three* crests, you had to become aware that there were more than one of these items. In other words you had to constitute 'pluralities' prior to assigning to them the number words 'two' and 'three' respectively. It is obvious that none of the individual troughs or crests tells the perceiver that it is part of a plurality. It is, again, the perceiver who, as Caramuel said, 'considers different things, each distinct in itself, and intentionally unites them in thought.'

Caramuel's statement describes the generative aspect of the procedure but does not specify how it works. Each of the troughs you saw, for instance, had to be a different, distinct thing, for otherwise you would have had no reason to say that there were two of them. However, they also had to be the same, in the sense that they were both troughs. A trough and a crest, taken together, would not yield a plurality. However, they could be used to form a plurality, if the sensory signals that led to their constitution were categorized differently — for example, as curves or deviations from the straight line.

This is a crucial observation. It shows that in order to construct a plurality, the perceiver must become aware of the fact that in the given perceptual context two (or more) unitary items have been categorized as belonging to one and the same class. In other words, the perceiver of the wave line must become aware of *repeating* a specific categorization. This introduces a higher level of operating into the procedure. Categorizing is an operation performed on the basis of sensory material. Recognizing the repetition of a categorization, however, is no longer dependent on specific sensory material, but requires the perceiver's reflection on his or her own operating.

Nothing but the reflection that recognizes the second (or further) categorization as a repetition of an earlier one can tell the perceiver that there is more than one unit of the same kind. The concept of plurality is one of the clearest examples of a 'reflective abstraction'. Together with the concept of unit it forms the basis of arithmetic and the mathematics involving numbers. In the pages that follow, I present a model that shows how these fundamental concepts could be constructed.

The Attentional Model

Assuming that the operation that creates unitary items is, indeed, independent of sensory signals, it would be tempting to suppose that it in some way involves motion. Piaget has long maintained that the perception of patterns is the result of active composition of sensory data by means of motion. Dividing a line into unitary sections might plausibly be achieved by movement alternating with pauses, and the same could be said in the case of the visual perception of items such as toothpicks lying on a table. By means of a minor additional hypothesis this idea could even be extended to situations where no direct perceptual scanning is involved.

There is a considerable body of evidence showing that figural composition can take place without any actual eye or body movement. Köhler (1951), Lashley (1951), Pritchard, Heron and Hebb (1960), and Zinchenko and Vergiles (1972), independently found that scanning of the visual field can take place when the field is stabilized on the retina and no eye movement can alter it. From the theoretical point of view, these findings are revolutionary. They indicate that a perceiver's attention can focus on one part of the visual field and shift its focus to another, without any corresponding change in the position of the sensory organ. This mobility of the focus of attention provides, on the one hand, an alternative to physical motion in the composition or integration of perceived patterns and, on the other, an active agent in the experiencer's organization of his or her experience.

Instead of tying the generation of unitary items to movements and pauses in the actual perceptual process (as suggested above), we can now attempt to account for it by the shifting and alternating of the focus of attention. This has the immediate advantage that it enables us to posit one and the same operational procedure regardless of what kind of item is being unitized. What I mean is this: from experience we know that we can conceive as one unit divide into several units, not only any array of perceptual signals, but also, for instance, last night's sleep or the rest of our lives. The same is the case with innumerable other conceptual constructs which, of their nature, are not dependent on perceptual signals to guide the unitizing operation.

The idea that the structure of certain abstract concepts could be interpreted as patterns of attention, was first proposed by Silvio Ceccato (1966). In the pages that follow I shall outline a possible application of that idea to numerical concepts. Attention, in this model, is conceived as a pulse-like activity that picks out, for further processing, some of the signals from the more or less continuous multitude of signals which the organism's nervous system supplies.³ That is to say, a single pulse or moment of attention can be, but need not be, focused on a particular signal. When it is unfocused it does not pick out particular signals, but this does not mean that there are no signals that could have been picked out. The unfocused moment merely creates a break in the process of composition. On the other hand, attention can focus on items that are not present as active sensorimotor signals, but as re-presentations of signals (or composites of them) that have been picked out at some prior occasion.

An Iteration of Pulses

It should be clear that I am using the word attention in a way that is somewhat different from ordinary usage. Expressions, such as focusing attention on a diagram or on the sunset, are used in situations where the speaker has posited, on the one hand, such things as a diagram or a sunset and, on the other hand, an organism that perceives those items — where 'perceiving' means to replicate as an internal representation something that is thought to be outside. This view is obviously incompatible with a constructivist orientation. Instead, I say that an organism focuses attention on signals in its nervous system. This implies that the organism must be able to operate on at least two levels. One, on which sensorimotor signals are generated and conveyed to other parts of the neural network, and a second level of *attentional* activity where focused pulses pick out particular sensorimotor signals, while unfocused pulses create discontinuities or intervals. To do this, the system needs some kind of memory where the results of attentional activity can be maintained in such a way that they, too, can subsequently become the object of attentional focusing.⁴ In short, in my model attention refers to a selective activity just as it does in ordinary usage. But the items that attention focuses on and selects are now items or events within the organism.

Given such a model that operates on several levels, one can attempt to map (as a crude approximation, to be sure) how an organism could come to have something like the concept of, say, an apple.

The partial definition of 'apple' I proposed earlier contained of a number of characteristics. Taste and smell, would be supplied by sensory signals, Shape, size, and texture would be combinations of visual, tactual, and proprioceptive (motor) signals. Weight would be tactual and proprioceptive; and the characteristic arrangement of skin, flesh, and core would probably involve colour and other visual as well as tactual signals. Assuming that the model organism were now to discover in its experience that these specific sensorimotor signals quite frequently occurred together and could, in certain contexts, be combined in an aggregate, the process of concept formation could be implemented through the simple extraction of those signals that are common to all (or at least most) of the occurrences.⁵ There is, however, one further condition: whatever the pattern of sensorimotor signals involved, it must be such that it constitutes a consecutive sequence of focused moments of attention. If it were not consecutive, but contained intervals of unfocused moments, it could not be categorized as a whole or unitary item. It is the two moments of unfocused attention at the beginning and the end of a sequence that provide the closure and cohesion of a unitary item. A mapping or diagram of the conceptual structure of a perceptual thing, such as an apple, could look like this:

$$O \left(\begin{array}{ccccc} I & I & I & \cdot \cdot \cdot & I \\ a & b & c & \cdot \cdot \cdot & n \end{array} \right) O$$

where 'O' designates unfocused moments of attention, 'I' focused moments, and 'a, b, c, . . . n' different sensorimotor signals that were individually picked out by consecutive focused moments of attention. This is a crude approximation because, as I mentioned above, even a relatively simple concept, such as apple, involves substructures in which sensory and motor elements are combined in specific characteristic ways. These substructures would have to be represented by parentheses or some other notational device. However, my concern here is not with the sensorimotor discrimination of different kinds of objects but rather with the feature they all must have in common if they are to be considered as units.

According to this model, a discrete unitary perceptual item is constituted by an attentional pattern that consists of an unfocused moment, an unspecified sequence of focused moments, and a terminal unfocused moment that closes it. In the suggested graphic notation, it would be represented by the sequence:

$$O \quad I \quad I \quad \cdot \quad \cdot \quad \cdot \quad I \quad O$$

The concept is the result of a 'reflective abstraction' that separated and retained an operational pattern from the sensorimotor material that provided the occasion for its constitution. In a further step of abstraction, the uninterrupted sequence of focused attentional pulses becomes fused and yields the generic attentional pattern of a *unit*:

$$O \quad I \quad O$$

This represents a wholly abstract entity, because it no longer matters what the central moment of attention was focused on or whether there was one or several.

I suggest that this is the conceptual pattern that Euclid had in mind when he wrote: 'A unit is that by virtue of which each of the things that exist is called one' (Euclid, Book VII). It was not a very helpful definition, because it did not say what the entity was 'by virtue of which' we call things 'one'. The attentional pattern provides a hypothetical model of that entity. It seems to fit the situations in which we construct units.

In the case of the four clocks in Caramuel's library, for instance, each clock could be perceived as striking in a different tone. In that case, there would be four different sensory signals and they would be considered as four single equivalent units, because each would be experienced as the same attentional pattern, namely O I O. The whole experience could be mapped as:

$$\underset{a}{O \ I \ O} \quad \underset{b}{O \ I \ O} \quad \underset{c}{O \ I \ O} \quad \underset{d}{O \ I \ O}$$

where a, b, c, d are the different sensory signals picked out by focused moments of attention. When only the attentional *pattern* is being considered and the sensory signals are disregarded, each of the strokes can be categorized as an instantiation of the concept of 'one'. A succession of such ones constitutes a 'plurality'. As there is no initial unfocused moment, nor a terminal one, that could serve as boundary and enclose them in a frame, the unit patterns remain individuals that are connected by nothing but their contiguity in experience.

This analysis provides the key to an ambiguity of which we are always more or less dimly aware: 'one' seems to refer to two concepts. Their difference becomes apparent when 'one' is opposed to 'many' and then to 'two', 'three', etc. The first opposition is the same as that between a singular and a plural or between unity and plurality. In the second, 'one' functions as number word, and there is no proper opposition, but merely the difference between one number and other numbers.

The Genesis of Plurality

Caramuel's insight that it is one thing to count, and quite another to consider several things contemporaneously, was uncannily correct. The mere repetition of the attentional pattern that creates unitary items is not counting but merely establishing a plurality. In order to count, Caramuel said, 'the intellect . . . considers different things, each distinct in itself, and intentionally unites them in thought.' A plurality is, indeed, made up of different items, each a discrete unit separated from the others by moments of unfocused attention (the terminal moment of the preceding item and the initial moment of the subsequent one).

The concept of plurality has another characteristic that is not often mentioned: it has neither end nor beginning. I have an example at hand. I take a look through the window at the end of my study: there is the road in the distance, and cars are passing. I use the plural of 'car' because I saw a plurality of them, namely more than one. I have no idea of a number. In order to give a reasonable answer to the question 'How many?', I should have to create an experiential boundary around the plurality. I should have to specify a length of time and count. The books on my desk also constitute a plurality, but there the experiential boundary is implied by the desk, and therefore I could count them at once (and would not be concerned with books on the shelves or in the bedroom).

There is a subtle alternative in bounding a plurality. On the one hand, it can be bounded by moments of attention focused on the experiential frame. In the case of the books, there is a moment's attention on the table, then a plurality of books as unitary items, and another moment focused on the table. This produces the conceptual structure expressed by 'the books on the table'. On the other hand a reflective abstraction can transform a plurality in a way that is analogous to the transformation of a perceptual item into the concept of unit. It drops out the sensorimotor material and constitutes the plurality as a unitary concept. This allows us to speak of a plurality as a discrete concept although the sequence of unitary items it comprises is itself unbounded.

In a certain respect this conceptual wrapping or packaging is similar to the way we manage the memories of past experiences in general. If you say to your companion: 'Remember our trip to Greece, last summer?', the reference to the trip opens a path to a host of remembered experiences, some of which may not immediately be available as re-presentations; and the reference to last summer locates this chain in a much larger array. There is a kind of nesting, and items can be nested even if, conceptually, they are not fully determined at the moment.

Another example is the use of verbs that designate an iterative action, such as 'to walk', 'to swim', 'to hammer'. These actions are characterized by the repetition of a specific procedural sequence. The user of the verb must know the sequence, but the beginning, duration, and end of the repetitions are left unspecified in the conceptual structure that constitutes the meaning of the verb (if they are relevant in a situation where the verb is used, they may be indicated by the context or other words). Thus, activities can be 'packaged' as unitary concepts, though there is no indication of a beginning or end. The same can be done with a plurality that has no beginning, no end, and therefore no numerosity. If we have a rule, i.e., an operational recipe, that governs how each of the component units is derived from the preceding one, we can package even a potentially infinite sequence of items and turn it into a unitary concept.[6]

The Abstract Concept of Number

As Peano remarked , Euclid's definition, 'A number is a multitude composed of units' (Euclid, Book VII), is insufficient. We can now see why. A plurality, too, is a 'multitude composed of units', but it does not involve the concept of number. Indeed, Euclid's definition would also fit a herd, a forest, a committee, and a stamp collection. Hence, an essential ingredient must be missing.

In my model, the transformation of a plurality into the kind of composite unit that can be considered a number, requires two further operations. The first is what I would call a 'conceptual iteration', the second is the activity of 'counting'.

They are frequently performed together. A simple, seemingly silly exercise may help to show what they are. Assume that you were asked how many lines of print there are in the paragraph following the one you are reading now. You move your glance to the indicated paragraph. From prior knowledge or as you move your glance, you categorize it as a bounded plurality of lines. You gather that this plurality is bounded, because the word 'paragraph' implies this and you can tell the boundaries from the indentations of the print. Next, you move your gaze up or down across the lines, and you coordinate with each one a number word of the conventional number word sequence, beginning with 'one'. Having reached the last line of the paragraph, the last number word used will tell you how many times you have repeated the compound of the two operations — it will indicate the *number* of lines.

Some of the indispensable features of this procedure have already been discussed. The unitary items of the plurality must be distinct from each other but considered the same in some respect. The plurality must be bounded, for if it were not, it would not be countable. The number word sequence must be known and strictly maintained, and it must start with 'one', The coordination of its elements and the unitary items to be counted, must be strictly one-to-one. All this has been said innumerable times, but it is worth repeating because descriptions of counting all too often neglect one aspect or another.

In addition, there is a feature that is crucial for the attentional model. When you were scanning up or down across the lines to be counted, you already knew the lines to be unitary items and did not have to reconstruct them perceptually as such. Instead you merely focused on the blackness or some other property of the printed lines and on the sequence line–interval–line–interval . . . etc. The intervals could have been of any kind, as long as they were not print. They simply supplied the occasion for moments of attention that were unfocused relative to those that were focused on the lines. This is the reason why I speak of 'iterating attention'.

What constitutes the abstract concept of *number* is the attentional pattern abstracted from the counting procedure. In this pattern it is irrelevant what the focused moments of attention are actually focused on. The salient features are: (1) the iteration of moments that are focused on *some* unitary items and attentional moments that are not; (2) that the iterated sequence itself is bounded by unfocused moments; and (3) that the focused moments are coordinated with number words. The pattern can be mapped as the following diagram:

$$O \; (O \; I \; O \; I \; O \; I \; \cdot \; \cdot \; \cdot \; O \; I \; O) \; O$$
$$\text{'1'} \quad \text{'2'} \quad \text{'3'} \cdot \; \cdot \; \cdot \quad \text{'n'}$$

According to Peano, you may remember, individual items have to be characterized by individual characteristics. In the diagram, the characteristic of each individual number is provided by the value of 'n'. A class, he said, is characterized by a common characteristic of its members. The general concept that comprises all individual numbers, therefore, is a further abstraction from the attentional pattern of individual numbers. The operation is analogous to the 'fusion' of focused attentional pulses in the generation of the abstract unit. In my notation it would be this:

$$O \quad (O \quad I \quad O) \quad O$$

The central 'I' would indicate a moment of attention that could be focused on *any* individual number that is considered the last in a counting action. This could explain why we have no difficulty in considering 'one' a number, in spite of the fact that it does not consist of a 'multitude of units' but of a single one. And, most important, in this unitary abstract concept of number all sensory material has dropped out.

The 'Pointing Power' of Symbols

The conceptual transformation of a plurality into the concept of number has intermediary steps that I described elsewhere (Glasersfeld, 1981a). Here, the simple counting exercise seemed sufficient to illustrate the main features. The symbol 'number' acquires its meaning from the fact that it *points* to the conceptual structure of attentional iteration in an actual or potential counting situation. In my discussion of reflection and abstraction (Chapter 5) I explained the pointing function of symbols. In the case of the concept of number, this function is crucial. Number words and all kinds of numerals point to specific instantiations of the number concept's attentional structure, but this does not entail that the indicated attentional iteration and the count have to be carried out. To understand the symbols, one merely has to know the required procedure and that it *could* be carried out.

In order to understand what is intended when we read the numeral '573' or '1001', or if someone mentions that the United States deficit has reached 'three trillion', we do not have to re-present to ourselves the indicated pluralities of units — we merely have to know the procedure that could produce and count them.

This, it seems to me, somewhat demystifies the otherwise astonishing fact that we can have and operate with the concept of an infinite number sequence. The answer lies in the fact that we can 'package' generative procedures and treat them as unitary conceptual entities. We do this every time we use an action verb such as 'to walk', and we can do it equally with abstract procedures. In the case of numbers, we know the process of attentional iteration and our system of number words is such that the counting procedure can be extended indefinitely. We therefore know that once these operations have been packaged as a unitary concept, they can very well be thought of as continuing endlessly inside their conceptual warpping.

Mathematical Certainty

The conceptual capability of packaging procedures also demystifies the eternal question about the certainty of mathematical knowledge. When I said at an earlier point (Chapter 2) that 2 + 2 = 4 is not questionable, I gave as the reason the agreement on the counting procedure and the number word sequence. However, there is another problem. The numerals (or number words) must be understood as indicating specific pluralities of units and one might object that, in practice, the perception of unitary items is often open to question. Yet, the results of the arithmetical operations with numbers are not considered questionable. In fact, they are as certain as the conclusion in a syllogism.

I have argued in the preceding sections that common non-mathematical activities, such as isolating objects in the visual or tactual field, coordinating operations while they are being carried out, and generating a line by an unchanging continuous movement of attention, form the experiential raw material that provides the thinking subject with opportunities to abstract elementary mathematical concepts. If one accepts this view, one is faced with the puzzling question how such obviously fallible actions can lead to the certainty that mathematical reasoning affords.

The model's answer to this puzzle lies in the fact that in the construction of the abstract concept of number all sensory material is eliminated. Although the numbers '1', '2', '3', and so on, were originally conceived with the help of experiential *things*, their sensory properties were dropped during the two steps of abstraction, first of units and then of units of units; and when we operate with abstract entities, we do not question that they are indeed abstract and no longer subject to the fallibility of sensory perception.

This is analogous to the certainty we attribute to the deductive procedure in a syllogism. If we write the traditional textbook syllogism with a first premise that we assume to be false — for instance, 'all men are immortal' — and then proceed with 'Socrates is a man', we get the obviously false conclusion that Socrates in *immortal*. Here the puzzle arises from the realization that this conclusion is just as certain and logically 'true' as its opposite, which we get when we start with the more plausible first premise that asserts the mortality of all human beings.

The puzzle disappears if it is made clear that the premises of a syllogism must be considered as 'hypotheses' and should be preceded by 'if'. Their factual relation to the experiential world is irrelevant for the formal functioning of logic. By considering them to be 'as if' propositions, we make sure that, for the time being and during the subsequent steps of the procedure, we are not going to question them. Hence the certainty of the conclusion springs from the fact that the situations specified by the premises are *posited* and, therefore, not to be questioned during the course of the procedure.

The analogy to the certainty of '2 + 2' in arithmetic lies in this: the symbol '2' stands for a conceptual structure composed of two abstract units, to which the number words 'one' and 'two' were assigned respectively. The symbol '+' requires that the units on its left be lined up with the units on its right and subjected to a new count. Since the standard number word sequence is fixed, and the items in the count are not questionable sensory things but abstract units, there is no way it could not end with 'four'.

Notes

1 Some of the ideas discussed in this chapter were first presented in Glasersfeld, 1981c.
2 The expression thinghood is intended to designate the separation of an item as a unit from the experiential field, much as in the realms of vision and art a figure is separated from the ground. This must not be confused with the concept of object permanence, a far more complex structure that involves both externalisation and re-presentation, neither of which is required in thinghood.
3 Ceccato's idea of the constitutive role of attention in the construction of concepts has recently been further elaborated by Vaccarino (1977, 1981, 1988) and Accame (1994).
4 Such a system of two or three levels is obviously still much too crude to account for most of the conceptual results a human organism can produce. Hypnosis suggests that things can be remembered even if they were not consciously experienced, and the work of Hilgard (1974) indicates that there are probably several levels of attentional activity that are relatively independent of one another.
5 With certain things there might be an obligatory order for some of the signals, in others it could be just a list. In the case of the wave line, for instance, 'crest' requires the sequence low–high–low, whereas 'trough' requires high–low–high.
6 This should be of interest to philosophers of mathematics who have been worrying about whether or not infinite progressions could be considered 'real'.

Chapter 10

To Encourage Students' Conceptual Constructing

During the last few years, the number of references to radical constructivism in the educational literature has increased in a startling fashion. It has made me very uneasy. If research programmes and schools announce that they have adopted the 'constructivist paradigm', innocent people are led to believe that there has been a breakthrough and that the adoption of constructivism will rescue education from whatever crisis it is thought to be in. This, of course, makes no sense — and, from my point of view, it is counter-productive. If such high expectations are raised, the backlash is bound to come before the few serious applications of the constructivist approach that are in progress will constitute a solid test. It takes a good many years to assess whether a novel attitude is actually helpful as an orientation for schools and teachers. Set theory was introduced a few decades ago with genuine hopes and fanfare, but turned out to be a flop as a teaching and learning device. There was, of course, an important difference. The people who recommended it, and those who adopted it, had a fairly clear idea of what it was. In the present vogue of constructivism this does not seem to be the case. Some of its advocates tout it as a panacea but would reject it if they became aware of its epistemological implications. At the other end of the scale, some of the critics jump to the conclusion that it denies reality, and therefore is a heresy they cannot fit into their orthodox metaphysical beliefs.

What Is Our Goal?

Because constructivism is a theory of knowing and cuts loose from traditional epistemology, its application to education requires first of all a clarification of what one intends to achieve. This raises a fundamental problem. Education, after all, is a 'political' enterprise. Its purpose, as I see it, is two-fold. On the one hand, students are to be empowered to think for themselves and without contradictions. On the other, the ways of acting and thinking that are at present judged the best, are to be perpetuated in the next generation.

Constructivism has no difficulty in accepting these premises, but it does not accept the usual justification of knowledge. In the traditional view, schools are seen as institutions that are to impart value-free, objective knowledge to students. For constructivists there is no such thing, because they see all knowledge to be instrumental. The first thing required, therefore, is that students be given the reasons *why* particular ways of acting and thinking are considered desirable. This entails explanations of the specific contexts in which the knowledge to be acquired is believed to work. Such explanations are profoundly shocking to those who believe in 'Truth for Truth's sake'.

The constructivist orientation is particularly distasteful to teachers (and students) in mathematics and the physical sciences whose conception of science has been shaped by scientistic myths in textbooks, television, and popular accounts of 'breakthroughs'. Students come into school with the preconception that science will tell them what the real world is like; and teachers fear that if they gave up the claim of objective truth they would lose their authority (see Désautels and Larochelle, 1989).

However, I am convinced that, in general, students will be more motivated to learn something, if they can see why it would be useful to know it. Most of the goals that determine the instrumental value of a piece of knowledge are not so arcane that students would not be capable of sharing them. This goes from meeting the prosaic material needs of everyday life to the generation of peace of mind on the abstract level of the individual's organization of experiential reality.

Unfortunately, the way the schooling system is set up (please note that for almost thirty years I have been working in the United States and know nothing of schooling systems elsewhere) has led to the widespread notion that one studies in order to pass exams, rather than to become more competent intellectually. This replaces the precious asset of knowledge with the paper money of certificates and degrees.

However, even if there were agreement about what should be learned and why, there would still be major problems about how it could best be taught. Indeed, the roots of the present crisis in education are many and diverse and even if a change of philosophy could suddenly be implemented, it would not bring about a cure at once. It takes time to modify habitual attitudes and expectations.

In any case, as radical constructivism holds that there is never only *one* right way, it could not produce a fixed teaching procedure. At best it may provide the negative half of a strategy. As I have often said, constructivism cannot tell teachers new things to do, but it may suggest why certain attitudes and procedures are fruitless or counter-productive; and it may point out opportunities for teachers to use their own spontaneous imagination.

There have been excellent teachers at all times, but they were often hampered because the methods they wanted to use did not fit the didactic conventions that governed schools. The constructivist orientation may do some good in this regard. It is a philosophy that offers a congenial theoretical basis for the development of imaginative teaching methods. It is in this spirit that I venture to voice my suggestions. They are not intended as directives but as encouragement.

Teaching Rather than Training

In order to adopt the constructivist way of thinking, some of the key concepts underlying educational practice have to be refashioned. The theoretical notions concerning the processes of communication and learning, the nature of information and knowledge, the interaction with others, and the phenomenon of motivation all change when they are seen from the constructivist perspective. Most of these changes were mentioned or implied in a general way in the preceding chapters. In what follows I shall try to pinpoint some of their effects in the context of education.

Teaching and training are two practices that differ in their methods and, as a consequence, have very different results. I have reiterated this many times. Only quite recently, however, did I discover that on this point, too, I am in agreement with Kant. While writing the brief historical survey of Chapter 2, I went back to Kant's collected works several times and came upon sections that I had never read before. In volume IX, the last containing writings published during his lifetime, I found his essay on pedagogy. It appeared in 1803, the year before he died.

> The human being can either be merely trained, broken in, mechanically instructed, or really enlightened. One trains dogs and horses, and one can also train human beings. Training, however, does little; what matters above all is that children learn to think. The aim should be the principles from which all actions spring. (Kant, 1803, *Werke*, vol ix, p.450)

In Kant's time, it seems, this was seen as an alternative. We have the behaviourist movement to thank for eliminating the path of rational enlightenment. By focusing exclusively on environmental stimuli and reinforcement, Behaviourism effectively obliterated the concern for thinking. Performance became the sole target. As a result, we still have tests that require students to do no more than remember what the teacher or the textbook has said. They test memory and rote learning, not understanding. Understanding, like mind and meaning, was considered a 'pre-scientific, mentalistic' fiction (see Skinner, 1971, pp.12–23).

From the constructivist point of view, the behaviourists' notions of 'stimulus' and 'reinforcement' are naive and misleading. The behaviourist movement, however, was not only enormously powerful a few decades ago, but its key notions are still alive and active in the minds of many educators. Hence, it may be useful to discuss our conceptual differences with regard to its two fundamental terms.

Environmental Stimuli

In the tradition of psychology, 'stimulus' refers to a percept that is expected to be followed by a 'response' of the perceiver. The terms originated in the study of reflexes, and the relation between them is tacitly assumed to be that of a cause and its effect. However, both cognitive psychology and cybernetics have shown that all the more interesting behaviours of living organisms cannot be reduced to the pattern of the reflex. The difference was made explicit in the model of the feedback loop. A percept does not trigger a response unless it shows a discrepancy with some *reference* that governs the organism's equilibrium.

Farmers, of course, never needed a scientific model to know this. They have always been aware of the fact that, although you can lead horses to the well, you cannot make them drink. It is not the external perception of water that causes them to be interested in water, but an internal one of thirst; and their thirst is essentially a subjective phenomenon to which only they have direct access.

The behaviourist dogma, however, holds that a scientific explanation can take into account only what is directly perceivable by an observer. On the one hand, this limitation leads to the programmatic neglect of all the internal reference values which, from our point of view, supply *reasons* for behaviour. These references are not only unobservable, they are also far from constant in each individual. (When I am idle or bored, the first ring of the telephone is a stimulus that propels me to answer; when I am immersed in work, I will let it ring for quite some time in the hope that it might stop.)

On the other hand, the exclusive focus on the observable imposes a misleading definition of what constitutes a stimulus. To assume that what observers isolate in their own perceptual fields as stimuli must be the same as what functions as stimulus for an observed organism, is a presumption based on the most naive form of realism. Animal psychologists, at least since Jakob von Uexküll and Georg Kriszat (1933), have become aware of important differences in the perceptual worlds of different species, and any human adult who interacts with another has opportunities to notice that in many situations the other perceives and attends to things that are different from those one attends to oneself.

In work with children or young students who are not yet accustomed to the perceptual and conceptual habits and constraints in a particular discipline, this discrepancy can be a serious stumbling block. All too frequently a 'fact' or a relation that seems perfectly obvious to the teacher is not even seen by the student.

Reinforcement

If an organism is in a state of perturbation (e.g., hunger) because internal sensory signals it receives indicate a deficit (e.g., lack of food) relative to the particular reference, it will be inclined to act in any way that reduces this deficit. Hence, if one sees to it that an animal is kept at 80 per cent of its normal body weight and thus in a constant state of acute hunger, it will be eager to repeat whatever behaviour the experimenter or the Skinner box *reinforces* with a bit of food. This principle was no doubt discovered by the first people who domesticated dogs and horses, although they probably did not methodically starve the animals, but merely retained control over their food.

No matter how it is acquired, the association of stimulus and reinforced response becomes a rather durable link and can be activated even in the absence of the original perturbation of hunger. The behaviourists refined the basic method by developing optimal reinforcement schedules and were thus able to design enormously effective training procedures.

More open-minded psychologists distinguish two kinds of reinforcement according to its origin. They call one extrinsic, the other intrinsic. Only the extrinsic kind is directly perceivable by an observer, but both have the effect of increasing the probability that the organism's preceding behaviour will be repeated. Because only the observable was considered 'scientific' by behaviourists, they excluded intrinsic reinforcement from their theory of learning.[1] This exclusion led to a programmatic disregard of conceptual learning, and the consequences for the methodology of teaching were devastating. A recent observer made the following assessment:

> School environments typically use a variety of reinforcements, such as praise, rewards, and grades. All of these are examples of extrinsic motivators, in that an activity is engaged in order to get the promised incentive, whether it be a star pasted to a school paper or a good grade on a report card. (Rieber, 1993, pp.205–6)

There is no question that this procedure works — it produces the repetition of the reinforced behaviour. Encouraged by this success, behaviourists launched the idea that they held the key to all learning. The reason why this claim is exorbitant is that extrinsic motivators do not motivate an effort to *understand*. Kant put his finger on it when he explained the main difference between training and teaching in the context of moral education.

> If one punishes a child when it does what is wrong, and rewards it when it does what is right, it will do what is good in order to be better off. (Kant, 1803, vol.IX, p.480)

Reinforcement certainly increases the frequency with which the behaviour is performed as response to the conditioned stimulus, but it does this without any consideration of the reason why the particular behaviour should be desirable in the given situation. Children who have only been trained to memorize '12 × 12 = 144', have no way of answering the question '12 × 13 = ?', because they have no conception of how numbers function. The reasoning that, because it is one more time 13, the answer must be 144 plus 13, is out of reach for them.

This is, of course, an exaggerated example. By the time children are drilled in the multiplication table, they may have independently acquired a notion of number that is a little closer to the abstract concept, because they have had occasion to use number words and counting in everyday situations where some mental operating with the symbols' meaning is required. Later, when they begin to understand how multiplication functions on the conceptual level, facility with the multiplication table will provide useful shortcuts in actual computations. My point, therefore, is not that training, memorization, and practice are useless. I merely want to stress the fact that rote learning does not lead to what Kant called 'enlightenment', namely an understanding of the operative principles that govern the entire problem area.

Training based on specific external reinforcements, be they social approval or some kind of prize, tends to set a spurious goal for the students. Whatever they are given as reward for a good performance becomes the reason for performing. This creates a temporary motivation to repeat the successful efforts, but it does not create the desire to learn more or to seek for themselves solutions to novel problem situations.

The motivation to master new problems is most likely to spring from having enjoyed the satisfaction of finding solutions to problems in the past. It is the excitement of glimpsing a possibility, working it out, and arriving at a result that passes whatever tests one can apply to it oneself. This is quite different from being praised because one's results are considered right by someone else. The insight *why* a result is right, understanding the logic in the way it was produced, gives the student a feeling of ability and competence that is far more empowering than any external reinforcement. This self-generated empowerment almost certainly engenders the desire for extension, the desire to experience it in a new context, and to enlarge the range of experiences that one can deal with satisfactorily. If students do not think their own way through problems and acquire the confidence that they *can* solve them, they can hardly be expected to be motivated to tackle more.

In my view, this consideration implies a fundamental ethical imperative: teachers should never fail to manifest the belief that students are capable of thinking. In this regard I am in superficial agreement with Socrates: I, too, believe that students 'have it in them' — but as a capability of construction, not as preformed ideas.

The Deceptive Character of Language

Owing to their mysterious ability to replicate actions seen in others, children can be helped, without the use of language, to learn to walk, to tie shoe laces, and to throw a ball.[2] The talent is invaluable in training. In teaching for understanding, however, language is an indispensable tool. Yet, few educators give much thought to how linguistic communication functions. Because language, by and large, works well in everyday situations, there is the tacit assumption that it must also work in the classroom. Consequently it is often assumed that students' failures to understand what is being taught must be due to other causes. No doubt, other causes often play their part, but the blind faith in the efficacy of language is probably the most frequent impediment to successful teaching.

I discussed some principles of linguistic communication in Chapter 7, and from my perspective they have direct consequences for the practice of teaching. If the meaning of the teacher's words and phrases has to be interpreted by the students in terms of their individual experiences, it is clear that the students' interpretations are unlikely to coincide with the meaning the teacher intends to convey. This indeterminacy is inherent in the communication system. It can, of course, be compounded by a student's lack of attention, but it is not caused by this. It springs from the way language is acquired.

The inherent looseness of language does indeed make teaching difficult, but it by no means makes it impossible. The difficulty is greatly reduced if the teacher keeps in mind that the words he or she uses have, for the listeners, associative links to their own experiential worlds and *not* to an independently existing reality that would be the same for all. Language does not convey knowledge but can very well constrain and orient the receiver's conceptual constructing. If teachers remain aware of this principle, they will constantly test the students' interpretations and not rest until the responses seem compatible.

When children enter school, they must learn new uses of language. This is not immediately obvious to them; nor are teachers always aware of the fact that the educational rituals of the particular discipline they are teaching differ from those of ordinary communication. The sociologist Erwin Goffman (1956) spoke of 'rules of conduct' that pattern social behaviour, although the actions that are guided by them are usually performed unthinkingly. Much of what teachers do in the classroom is guided by such rules of conduct; and the way they use language and initiate linguistic interactions tacitly presupposes the knowledge of patterns and rules that are commonplace in an adult educator's conceptual world. They are second nature to the teacher but not to the child novice. Some of these patterns are counter-productive from the constructivist point of view because they were designed to facilitate training and tend to discourage questions, conversation, and individual reflection.

A tacit 'rule of conduct' is presupposed when problem solving is introduced as a didactic tool. It concerns the students' attitude rather than their behaviour. For adults, the mere proposition of a problem is sometimes enough to capture their attention and to start them working towards a solution. But this is by no means a general rule, and to assume that it will automatically work in the case of students, is hardly justified. If there are many things one would like to do, and most of the time one is prevented from doing them — which is very often the position of students — only problems that are congenial in some way are likely to trigger sufficient interest. This greatly limits the teacher's choice. In arithmetic, for instance, textbooks are often little help in this regard.

Problem solving is undoubtedly a powerful educational tool. However, I would suggest that its power greatly increases if the students come to see it as *fun*. I first saw this during my visits to the Purdue Project a few years ago (see Cobb, 1989; Cobb and Bauersfeld, 1993; Wood *et al.*, 1993). To witness children collecting around their teacher at the end of a 2nd-grade math class, and to hear them ask for more 'problems', was a revelation. How was it achieved?

This is a delicate question. Much depends on the teacher's sensitivity and willingness to go along with an individual student's way of thinking and, whenever possible, to involve the whole class in following and discussing the particular itinerary. The choice of task, of course is crucial and requires the teacher to use imagination rather than routine. A lesson can be started by letting a child recount an experience of his or her own that involves numbers. Usually it is not too difficult, then, to splice an appropriate 'problem' into the recounted story and thus to create some interest in the solution.

One of the secrets is to take the dreariness out of an obligatory occupation and to make it feel more like a deliberately chosen form of play. This is easy to say, and it may sound glib, because one can give no recipes, no standard procedures to achieve it. Yet, some teachers are able to do it. They develop a style that inspires relaxation and enjoyment because they feel at home with the subject matter and do not find the activity boring. Teaching, it has been observed, is an art.

The Orienting Function

This term was coined by Humberto Maturana (1970a) for the function of language in general. I want to illustrate by means of a rather crude metaphor the meaning it has for me in the context of teaching. Let there be no misunderstanding, the metaphor is intended to illuminate the dynamics of the situation, not the character of the subjects involved.

When a farmer has to drive a few heads of cattle along one of those small country roads flanked by hedges that have openings every now and then, the task is practically impossible if he has no helper. He has to stay behind the animals to keep them going, and when the first cow spots an opening in the hedge, it inevitably turns into the field. The others follow, and the farmer then has to run into the field to drive them back through the gap. This is difficult enough, but what makes the situation desperate is that the cows, forced back on the road, always turn into the direction from which they came.[3] It is a no-win scenario and no farmer would undertake such a trip without bringing along at least an obedient dog. This makes all the difference. Whenever the farmer spots a gap in the hedge ahead, he sends the dog to block it — and the problem does not arise. Note that the dog does not drive the cattle, it merely provides an additional constraint for their movement. It is the farmer who has to keep them moving. In this scenario, the dog has a function that is similar to an important use of language in the classroom.

The teacher cannot tell students what concepts to construct or how to construct them, but by a judicious use of language they can be prevented from constructing in directions which the teacher considers futile but which, as he knows from experience, are likely to be tried. As in the farmer's case, it is the teacher who has to provide the motivation to keep going, and although his language cannot determine the students' conceptual constructing, it can set up constraints that orient them in a particular direction.

Perceptual Materials

Not unlike is the role I would ascribe to many of the 'teaching aids', the perceptual props and arrangements that are used in classrooms, from number blocks and rods to experimental demonstrations in the physics lab. All too often teachers seem to be convinced that the abstract concepts and relations they are trying to convey are plainly visible in the physical material they are displaying. Mathematics teachers tend to forget that their approach to numerical symbols is governed by habits they did not acquire from one day to another. Similarly, science teachers tend to forget that their own way of looking and seeing has been conditioned by years of familiarity with the physicist's theory of motion or electricity, and that the particular patterns of 'conceptual conduct' that have become integral parts of their picture of the world are neither obvious nor God-given.

From the constructivist perspective, concepts are not inherent in things but have to be individually built up by reflective abstraction; and reflective abstraction is not a matter of looking closely but of operating mentally in a way that happens to be compatible with the perceptual material at hand. Hence, the physical materials are indeed useful, but they must be seen as providing opportunities to reflect and abstract, not as evident manifestations of the desired concepts. Cuisenaire rods, for instance are not *embodiments* of numbers, but their physical properties are such that they invite the construction of units and attentional iteration.

Individual students often make abstractions from the presented perceptual material that are quite different from those the teacher intends, to whom the material seems unambiguous. The seminal work on the relations of language and arithmetic by Hermine Sinclair (summarized in Sinclair, 1990), Les Steffe's work (1984, 1991) on teaching elementary arithmetic, as well as studies by Kamii and Joseph (1989) on the concepts of young children, and John Clement's (1983, 1993) reports on physics students at high-school and university levels, have consistently shown this. Hence it seems essential to provide a class with a variety of perceptual situations that can all be seen as instantiation of the conceptual construction the teacher wants to induce. In the search for possible analogies the teacher can foster discussion and orient the students' perspective without curbing their conceptual constructing by *telling* them what he considers right.

To sum up, both language and perceptual materials can provide experiential situations that may be conducive to reflections and abstractions a teacher wants to generate, but they are merely occasions, not causes. The students' concepts are determined by what they, as individual perceivers, come to abstract (empirical abstractions from their sensations, and reflective abstractions from the operations they themselves carry out in the process).

Students would be less likely to develop an aversion to mathematics and logical thinking if they were given the opportunity to grasp early on, that what they are expected to learn concerns mental operations and abstractions, rather than the actions and objects of the everyday world. There are many opportunities to do this, but all too often they are missed because the teacher feels obliged to convey what counts as accepted knowledge, rather than help students to build it up for themselves.

A Geometric Point

I have a vivid memory of how our teacher started us off in geometry. Chalk in hand, he made a small circular splotch on the blackboard and said, 'This is a point'. He hesitated for a moment, looked at the splotch once more, and added, 'Well, it isn't really a point, because a point has no extension.'

Then he went on to lines and other basic notions of geometry. We were left uneasy. We thought of grains of sand or specks of dust in the sunlight, but realized that, small though they were, they still had some extension. So, what was a point?

The question was buried in our struggle to keep up with the lessons, but it was not forgotten. It smouldered unresolved under whatever constructs came to cover it, and did not go away. In the course of the next few years it was joined by some other bubbles of uneasiness. When we came to infinite progressions, limits, and calculus, we were tacitly expected to think that there was a logically smooth transition from very small to nothing. We were told that Zeno's story of Achilles and the tortoise was a playful paradox, an oddity that did not really matter.

I did not like it, but I had decided to love mathematics anyway. Some of my schoolmates, however, concluded that mathematics was a silly game. Given the way some of it was presented, their reaction was not unjustified.

In retrospect, decades later, I realized that there had been quite a few occasions where the teacher could have resolved all those perplexing questions by one explanation. Shortly after the point episode in that geometry class, the teacher introduced the term 'equilateral triangle'. It was in the days when wooden rulers and triangles were used to draw on the blackboard. The teacher picked up one of these contraptions and showed it to the class. 'This is an equilateral triangle because its three sides have the same length'. As he was holding it up, he noticed that one of the corners was broken off. 'It's a little damaged', he said, 'it *would* be an equilateral triangle, if you imagine the missing corner.'[4] He missed a most appropriate occasion to explain that all the elements of geometry, from the point and the line to conic sections and regular bodies, have to be imagined. He could have explained that the points, lines, and perfect triangles of geometry are fictions that cannot be found in the sensorimotor world, because they are concepts rather than things. He could have told us that, no matter how exactly a physical triangle is machined, it is clear that, if one raises the standard of precision, its sides will be found to be not quite straight and their length not quite what it was supposed to be. He could have gone on to explain that mathematics — and indeed science in general — is not intended to describe reality but to provide a system for us to organize experience. I do not think that many students would be unable to understand this — and once it was understood, the domains of mathematics and science would seem a little more congenial.

Zeno's story of Achilles and the tortoise could serve as a powerful didactic tool. It would not be difficult to explain that if one thinks of Achilles and the tortoise as physical objects, the distance between them has to shrink only until it is less than the reach of Achilles' arm, and the tortoise will be caught. If, however, one thinks of them as two distinct geometrical points, one can conceive of more such extensionless points between them, no matter how often the interval has been halved, and the point called Achilles will never catch up with the point called tortoise. As in many other cases, the paradox disappears when the hidden conceptual incompatibility that generated it is brought to the surface.

The Need to Infer Students' Thinking

The fundamental principle from which most of my suggestions for the practice of teaching derive is that concepts and conceptual relations are mental structures that cannot be passed from one mind to another. Concepts have to be built up individually by each learner, yet teachers have the task of orienting the students' constructive process. Clearly it is easier to orient students towards a particular area of conceptual construction if one has some idea of the conceptual structures they are using at present. In other words, in order to modify students' thinking, the teacher needs a model of how the student thinks. Because one can never get into the heads of others, these models always remain conjectural (Glasersfeld and Steffe, 1991).

The teacher's assessment of a student's conceptual structures does not have to be a blind conjecture. If one starts from the assumption that students generally try to make sense of their experience, it is usually possible to get some idea of how they think. The more experience with learners a teacher has gathered, the better the chance to make an educated guess about what a particular student's thinking might be and to hypothesize what Vygotsky aptly called 'the zone of proximal development'. Sensitive teachers will treat their initial model of a student like a weather forecast: generally useful, though no better than approximate. It is only after working with a particular student for considerable time, that a teacher may gain confidence in his or her conceptual portrait of that individual. Needless to say, protracted experience with many students leads to plausible generalizations, but Les Steffe's painstaking microanalyses have shown that, even in the first grades, some individuals produce wholly unexpected inventions.

Research in physics education, as it was carried out by Andy diSessa, Rosalind Driver, John Clement, and a good many others, has shown that students have a small number of theories about the motion of cars, projectiles, and the various balls used in games. Although these preconceptions are on the whole incompatible with the physicists' explanation of the phenomena concerned, they contain, as Jim Minstrel (1992) documented, elements that are 'correct' and serve the students quite well in their daily lives. Telling them that they have to change their ideas because they are not 'true', may create obedient lip service but does not generate understanding.

Accommodation usually does not take place as long as a scheme produces the expected or desired result. Change may occur when a scheme fails or when a contradiction with another successful scheme surfaces. Yet, even among outstanding physicists, it is not the case that one single failure of an established theory would prompt them to relinquish it. As Kuhn (1962) has shown, 'normal science' within a given paradigm continues for quite some time in spite of the appearance of anomalies that put the paradigm in question. It is therefore rather naive to expect that one demonstration in class will induce students to give up a 'misconception' which they have found useful in their ordinary lives.

Students' unorthodox conceptions, much like mistakes they make in their attempts to solve problems, are among the clues from which the teacher can infer aspects of their actual conceptual network. More revealing still is what they say when asked to explain how they conceptualize a given situation and what general rules or 'laws' they apply to it. If the teacher at once reacts by saying that their ideas are wrong and tells them what is considered 'right', the students may indeed adopt the suggestion, but the reason why it is considered better may not be understood. It would seem more efficient to present the students with situations where the lay theory they have been using does not work. The motive to look for a more successful theory may then arise from their own perspective.

Help Rather than Instruction

When students are driven by their own interest to investigate and conceptually grasp a situation, the conceptual changes they are making during the process of reflection will be far more solid than if they were imposed by a teacher. Jere Confrey recently cited an excellent example.

> We have repeatedly witnessed situations in which teachers have overlooked opportunities to explore rich student ideas. At other times, they have unintentionally distorted students' statements to fit with their own understanding of the content. To illustrate the potential for this kind of silencing, consider the following story. In discussions about pyramids, children debated intensely whether the point on the top of the pyramid should be counted as a corner in a chart of sides, edges, and corners. Over time, many alternative views of corners and points emerged: a corner is where three faces meet; a corner is on the inside, and a point is on the outside; a point is not a corner because it can have more than three sides meeting; a point is not a corner because it does not touch the base. The teacher who guided this discussion, 'attempted to follow the students' thinking, asking questions designed to challenge and clarify their definitions [Russell and Corwin 1991, p.180]. Now suppose instead that she had quickly introduced the term *vertex* to include both corners and points. The discussion would have been prematurely terminated, and an opportunity for children to understand the role of definition and distinction would have been lost. (Confrey, 1993, pp.306–7)

A crucial aspect of the reported episode is the children's conversation. If we start with the notion that concepts can be abstracted and shaped only by the acting subject's reflection upon an experiential situation and the mental operations it provokes, we soon come to realize that talking about the situation is conducive to reflection. In order to describe verbally what we are perceiving, doing, or thinking, we have to distinguish and characterize the items and relations we are using. This often focuses attention on features of our construction that had remained unnoticed, and it is not at all uncommon that one of these features, when put into words, leads us to realize that some conclusion we had drawn from the situation is not tenable. (Any writer of a paper on work in progress knows this only too well.)

In this regard, I would reiterate what I wrote in my introduction to a special issue of *Educational Studies in Mathematics*:

> To engender reflective talk requires an attitude of openness and curiosity on the part of the teacher, a will to 'listen to the student' . . . it is one of the primary duties of the teacher to create an atmosphere in the classroom that not only allows but is also conducive to con-

versation, both between student and teacher and among students. (Glasersfeld, 1992c, pp.443–4)

Jack Lochhead, working with undergraduates in physics courses, frequently gave them problems they had not encountered in the textbook but should have known how to approach, given the knowledge of physics they were supposed to have acquired. He let them work at the blackboard and encouraged them to explain what they were doing and why they were doing it. The reports of these sessions (e.g., Lochhead, 1988) would be instructive reading for most teachers. They demonstrate not only how the concepts and laws of physics are often misunderstood or misinterpreted, but also how powerful a learning experience it is for students to discover for themselves that what they are doing and describing makes no sense. Such moments of self-generated crisis are infinitely more conducive to conceptual accommodation than any external criticism. They are moments in which the teacher may become a most effective helper, not by showing the 'right' way, but by drawing attention to a neglected or counter-productive factor in the student's procedure. Teaching, as Gordon Pask often says (e.g., 1961, p.89), must be a form of conversation.

In order to help, however, the teacher must have some idea of the kind of conceptual change that would, at a particular point, constitute an advance for a particular student. Recently, much has been written about 'higher order processes' and 'metacognition', but the level of discussion is all too often so theoretical that it is difficult to gather what could be done in practice. Teachers should not be expected to be experts in philosophy or semantics, and what is needed, therefore, is a down-to-earth approach. I received a paper the other day, that makes what seems to me a valid step in that direction: it defines different types of conceptual change and gives examples.

1   *Differentiation*, wherein new concepts emerge from existing, more general concepts — for example, velocity and acceleration emerging from generic ideas of motion . . .
2   *Class extension*, wherein existing concepts considered different are found to be cases of one subsuming concept — for example, rest and constant velocity coming to be viewed as equivalent from the Newtonian point of view.
3   *Re-conceptualisation*, wherein a significant change in the nature of and relationship between concepts occur — for example, the change from 'force implies motion' to 'force implies acceleration' . . .
    (Dykstra, Boyle and Monarch, 1992, p.637)

A catalogue of such patterns may give the teacher a better chance to make an educated guess about the student's 'zone of proximal development'. What Dykstra calls 're-conceptualization' clearly involves reflection. I would suggest that mapping items such as motion, velocity, and acceleration in terms of attentional frames (see Chapter 4) would induce students to reflect and help them to form new abstractions.

Fostering Reflection

Putting students into groups of two or three and designating the one the teacher considers the 'weakest' to report on their results at the end of the session, seems an excellent strategy. It compels them to explain their thoughts to one another and this has several advantages: on the one hand, verbalization requires reflection (upon one's own thoughts as well as upon what the others are saying) and, on the other, students tend to listen more openly and with more interest to their fellow students than to the teacher.

Paul Cobb, has used this method with great success for teaching arithmetic in elementary school classes (Cobb *et al.*, 1993; Cobb and Bauersfeld, in press). Above all, he and his colleagues have demonstrated that one of the most frequent objections to group work is not valid. People involved in education say that if a substantial part of lessons is devoted to groups working more or less on their own, the curriculum cannot be covered and the tests at the end of the year turn out to be a disaster. Cobb's reports have shown that this is an exaggeration even if only the first year of his experiment is considered. After three years, the children in his project not only attained better-than-average test scores in arithmetic but also in other areas. They had acquired a better way of learning (Wood *et al.*, 1993).

The flaw in the traditional assessment of student progress is that progress is assumed to manifest itself as a linear sequence of advances in competence, each of which can be shown in a test the moment it has taken place. This view is inadequate, even in the context of acquiring competence in sensorimotor skills such as playing tennis or skiing. Almost without exception, observable or testable progress is preceded by small steps of internal reorganization that remain hidden to the observer. Then, at a certain moment, they produce a perceivable change in the learner's performance. When we are dealing with conceptual understanding, where the way changes come about is largely a matter of inference, this consideration is all the more relevant. Experience has shown that there are usually more or less long periods of *latency* where change is not observable at all. In retrospect from a later point in the development, however, the conclusion that a number of internal changes must have taken place becomes inescapable.

As a result of listening for several years to the conversations of groups of two or three students in arithmetic classes, Susan Pirie and Tom Kieren (1989) were able to formulate a detailed theoretical model of conceptual change. It hypothesizes a cyclical pattern of well-defined steps that allows the teacher to see patterns in what the students do and say. As such, it should prove a useful tool for the systemic organization of observations. From the constructivist point of view, that is the purpose of theories.

The Secret of 'Social' Interaction

What we see others do, and what we hear them say, inevitably affects what we do and say ourselves. More important still, it reflects upon our thinking. If one takes seriously the idea that the others we experience are the others we construct, it follows that whenever they prove incompatible with our model of them, this generates a perturbation of the ideas we used to build up the model. These ideas are *our* ideas, and when they are perturbed by constraints, we may be driven to an accommodation. Socially oriented constructivists speak of 'the negotiation of meaning and knowledge'. This is an apt description of the procedure because, as a rule, it takes a sequence of small reciprocal accommodations to establish a modicum of compatibility.

Teachers who start out with the conviction that there is a fixed body of knowledge that has to be instilled into the students, are unlikely to see their activity as a form of negotiation. Yet, those who have a record of effective teaching and have begun to examine what it was that made them effective, are no doubt aware of the fact that any given piece of knowledge may be approached and then *seen* differently by individual learners.

For radical constructivism, the crucial aspect of the 'negotiating' procedure is that its results — the accommodated knowledge — is still a subjective construction, no matter how mutually compatible the knowledge of the negotiators may have become in the process. I know no simpler and more lucid formulation of the basic constructivist view than the one Heinrich Bauersfeld gave when he explained the role of negotiation in the generation of knowledge a few years ago:

> Altogether, the subjective structures of knowledge, therefore, are subjective constructions functioning as viable models, which have been formed through adaptations to the resistance of 'the world' and through negotiations in social interactions. (Bauersfeld, 1988, p.39)

From the *rational* perspective that I have tried to illustrate and maintain throughout this book, there is no functional difference between the constraints the builder of action schemes and conceptual structures meets in the form of physical obstacles and the resistance manifested in interactions with people. The movement that calls itself 'social constructionism' disagrees with this view. Language and social interaction, they claim, provide more direct means for the *Sharing* of knowledge (e.g., Gergen and Gergen, 1991, p.78). As a constructivist, I would never say that they are wrong, but I would ask that they present a plausible model of *how* such sharing of meaning, and the *collective* generation of knowledge in language, can take place.

A Final Point

Much of what has to be taught in mathematics and the sciences is fairly remote from the students' daily lives and interests. The motivation to learn can spring from a variety of sources, but they are rarely ready and flowing when the teacher begins. They have to be tapped. Creating a plausible link between the subject matter and the students' field of experience, is a good way, but not always possible. Another — that of course makes much greater demands on the teacher — is the display of honest enthusiasm for the topic and its problems. Students have a keen eye for fake enthusiasm, and teachers who feel that their authority lies in knowing all the answers have little chance of awakening genuine curiosity in their students. In my view, a teacher should always welcome the opportunity to work with students on a problem to which he or she does *not* know the answer. On such occasions a great deal of authority can be gained, not by pushing a better way to solve the problem, but by using arguments within the students' horizon that show why some of their suggestions are inappropriate and unlikely to succeed.

I say this (like so much in this book) on the basis of a high-school experience. When we came to number theory, we had a true mathematician as teacher. The passion with which he tried to show us why certain proofs were elegant and others correct but tedious, made us eager to see the distinction. One day he mentioned that he spent most of his free time working on Fermat's 'Last Theorem'.

'It looks so simple', he said, 'who knows — may be one of you can find the proof.' We spent the weekend sweating over it, rather than skiing. During the next lesson he looked at our attempts and gently suggested why he thought they went into unlikely directions. From this experience we learned something about *learning* that the curriculum did not supply.

To sum up, what radical constructivism may suggest to educators is this: the art of teaching has little to do with the traffic of knowledge, its fundamental purpose must be to foster the art of learning.

Notes

1   Nevertheless they used it as an explanatory device (see Skinner, 1971, p.107).
2   Psychologists usually speak of imitation as a commonplace phenomenon that requires no explanation. Yet, as far as I know, there is no model to explain how the visual impression of an action could be translated into a motor pattern.
3   I experienced this as a novice farmer in Ireland, but I believe the scenario would be the same in parts of England and wherever country roads are flanked by hedges or fences.
4   I used this anecdote at the International Workshop on Physics Learning in Bremen (see Glasersfeld, 1992b).

References

ACCAME, F. (1994) *L'individuazione e la designazione dell'attività mentale*, (The isolation and specification of mental activity), Rome, Editrice Espansione.

ARCHIVES JEAN PIAGET, (Ed) (1989) *Bibliographie Jean Piaget*, Geneva: Fondation, Archives J. Piaget.

ASHBY, W.R. (1952) *Design for a Brain*, New York, Wiley.

BACON, F. (1623) 'The dignity and advancement of learning', in DEVEY, J. (Ed) (1881) *Works of Lord Bacon*, London, Bell.

BARTHES, R. (1987) *Criticism and Truth*, Minneapolis, University of Minnesota Press (French original, 1962.)

BATESON, G. (1972a) 'Metalogue: What is an instinct?', in BATESON, G. (1972b) pp.38–58.

BATESON, G. (1972b) *Steps to an Ecology of Mind*, New York, Ballantine.

BAUERSFELD, H. (1988) 'Interaction, construction, and knowledge', in GROUWS, D.A. and COONEY, T.J. (Eds) *Perspective on Research on Effective Mathematics Teaching*, Reston, Virginia, National Council of Teachers of Mathematics.

BELLONI, L. (Ed) (1975) *Opere scelte di E. Torricelli* (Selected works of E. Torricelli), Turin, Unione Tipografico-Editrice Torinese.

BENTHAM, J. (1770ff) *Theory of Fictions*, in Ogden, C.K. (1956).

BERGER, P.L. and LUCKMANN, T. (1967) *The Social Construction of Reality*, Garden City, New York, Anchor Books.

BERKELEY, G. (1706–8) 'Philosophical Commentaries' (also known as *Commonplace Book*) in LUCE, A.A. and JESSOP, T.E. (Eds) (1950) vol.I.

BERKELEY, G. (1709) 'An essay towards a new theory of vision', in LUCE, A.A. and JESSOP, T.E. (Eds) (1950) vol.I.

BERKELEY, G. (1710) 'A treatise concerning the principles of human knowledge, in LUCE, A.A. and JESSOP, T.E. (Eds) (1950) vol.II.

BERKELEY, G. (1732) 'Alcyphron', in LUCE, A.A. and JESSOP, T.E. (Eds) (1950) vol.III.

BERNAL, J.D. (1971) *Science in History*, Vol.1–4, Cambridge, Massachusetts, M.I.T.Press (first published, 1954.)

BETH, E.W. and PIAGET, J. (1961) *Épistémologie mathématique et psychologie*, Paris, Presses Universitaires de France.

BICKHARD, M.H. and RICHIE, D.M. (1983) *On the Nature of Representation*, New York, Praeger.

References

BOGDANOV, A. (1909) (pseudonym: N. Verner) 'Nauka i filosofia' (Science and philosophy), in *Òcerki filosofii kollektivisma* (Essays on the philosophy of collectivism), St.Petersburg.

BOWER, T.G.R. (1974) *Development in Infancy*, San Francisco, Freeman and Co.

BRIDGMAN, P.W. (1927) *The Logic of Modern Physics*, New York, Macmillan.

BRIDGMAN, P.W. (1936) *The Nature of Physical Theory*, Princeton University Press (reprinted by Wiley Science Editions, New York, 1964).

BRIDGMAN, P.W. (1961) *The Way Things Are*, New York, Viking-Compass.

BRINGUIER, J-C. (1977) *Conversations libres avec Jean Piaget*, Paris, Robert Laffont.

BRONOWSKI, J. (1978) *The Origins of Knowledge and Imagination*, New Haven/London, Yale University Press.

BROUGHTON, J.M. (1981) 'The genetic psychology of James Mark Baldwin', *American Psychologist*, 36, 4, pp.396–407.

BROUWER, L.E.J. (1949) 'Consciousness, philosophy, and mathematics', *Proceedings of the 10th International Congress of Philosophy*, Vol.I, part 2 (1235–49) Amsterdam, North Holland Publishing Co.

CAMPBELL, D.T. (1974) 'Evolutionary epistemology', in SCHILPP, P.A. (Ed) *The Philosophy of Karl Popper*, LaSalle, Illinois, Open Court.

CARAMUEL, J. (1670) *Mathesis biceps, Meditatio prooemialis* (Binary arithmetic; Introductory meditation), Campania, Officina Episcopali (Italian translation by C. Oliva; Vigevano: Accademia Tiberina, 1977).

CECCATO, S. (1949) 'Il Teocono o "della via che porta alla verità"' ('Theocogno', or of the path that leads to truth), *Methodos*, 1, 1, pp.34–54.

CECCATO, S. (Ed) (1960) *Linguistic Analysis and Programming for Mechanical Translation*, Milan, Feltrinelli and New York, Gordon and Breach.

CECCATO, S. (1966) *Un tecnico fra i filosofi*, (two vols.), Padua, Marsilio.

CECCATO, S. and ZONTA, B. (1980) *Linguaggio, consapevolezza, pensiero* (Language, awareness, and thought), Milan, Italy, Feltrinelli.

CELLÉRIER, G., PAPERT, S. and VOYAT, G. (1968) *Cybernétique et épistémologie*, Paris, Presses Universitaires de France.

CHOMSKY, N. (1986) *Knowledge of Language*, New York, Praeger.

CLEMENT, J. (1983) 'Students preconceptions in introductory mechanics', *American Journal of Physics*, 50, 1, pp.66–71.

CLEMENT, J. (1993) 'Using bridging analogies and anchoring intuitions to deal with students' preconceptions in physics', *Journal of Research in Science Teaching*, 30, 10, pp.1241–57.

COBB, P. (1989) 'Experiential, cognitive, and anthropological perspectives in mathematics education', *For the Learning of Mathematics*, 9, 2, pp. 32–42.

COBB, P. and BAUERSFELD, H. (in press) *Emergence of Mathematical Meaning: Interaction in Classroom Cultures*, Hillsdale, New Jersey, Lawrence Erlbaum.

COBB P., Wood, T. and YACKEL, E. (1993) 'Discourse mathematical thinking, and classroom practice', in FORMAN, E., MINICK, N. and STONE, A. (Eds) *Contexts for Learning. Sociocultural Dynamics in Children's Development*, New York, Oxford University Press, pp.91–119.

CONANT, R. (Ed. 1981) *Mechanisms of Intelligence: Ross Ashby's Writings on Cybernetics*, Seaside, California, Intersystems Publications.

CONFREY, J. (1993) 'Learning to see children's mathematics: Crucial challenges in constructivist reform', in TOBIN, K. (Ed) *The Practice of Constructivism in Science Education*, Washington, AAAS Press, pp.299–321.

CRAIK, K.J.W. (1966) *The Nature of Psychology*, Cambridge, Cambridge University Press.

DÉSAUTELS, J. and LAROCHELLE, M. (1989) *Qu'est-ce que le savoir scientifique?* (What is scientific knowledge?), Québec, Presses de l'Université Laval.

DIELS, H. (1957) *Die Fragmente der Vorsokratiker*, Hamburg, Rowohlt.

DYKSTRA, D.I., BOYLE, C.F. and MONARCH, I.A. (1992) 'Studying conceptual change in learning physics', *Science Education*, 76, 6, pp.615–52.

EINSTEIN, A. (1954) 'Physics and reality', *Ideas and Opinions*, Bonanza Books, New York (First published in Journal of the Franklin Institute, 1936, 221, No.3.)

EUCLID (ca. 250 BC) 'Book VII', in HEATH, T.L. (Transl. and Ed) (1926) *The thirteen books of Euclid's elements* (3 vols.), Cambridge, The University Press.

EXNER, F. (1919) *Vorlesungen über die physikalischen Grundlagen der Naturwissenschaften*, Vienna, Deuticke.

FEYERABEND, P. (1975) *Against Method*, London, NLB, Atlantic Highlands, Humanities Press.

FIREMAN, G. and KOSE, G. (1990) 'Piaget, Vygotsky, and the development of consciousness', *The Genetic Epistemologist*, 18, 2, pp.17–23.

FOERSTER, H.VON (1965) 'Memory without record', in KIMBLE, D.P. (Ed) *The Anatomy of Memory*, Palo Alto, California, Science and Behavior Books (Reprinted in Foerster, 1981).

FOERSTER, H.VON (1973) 'On constructing a reality', in PREISER, F.E. (Ed) *Environmental Design Research*, Stroudsburg, Dowden, Hutchinson, and Ross (reprinted in Foerster, 1981), pp.35–46.

FOERSTER, H.VON (1981) *Observing Systems*, Seaside, California, Intersystems Publications.

FRASER, A.C. (Ed) (1959) *An Essay Concerning Human Understanding by John Locke* (Complete and unabridged edition, collated and annotated by A.C. Fraser), New York, Dover.

FREUD, S. (1930) *Die Traumdeutung* (The interpretation of dreams), Leipzig/Vienna, Franz Deuticke, 8th edition.

GALLUP, G.G. (1977) 'Self-recognition in primates', *American Psychologist*, 32, pp.329–38.

GERGEN, K.J. and GERGEN, M.M. (1991) 'Toward reflexive methodologies', in STEIER, F. (Ed) *Research and Reflexivity*, London, Sage Publications, pp.76–95.

GLASERSFELD, E.VON (1965) 'An approach to the semantics of prepositions', in JOSSELSON, H. (Ed) *Proceedings of the Las Vegas Conference on Computer-related Semantic Analysis*, Detroit, Wayne State University.

GLASERSFELD, E.VON (1974) 'Signs, communication, and language', *Journal of Human Evolution*, 3, pp.465–74.

GLASERSFELD, E.VON (1976a) 'The development of language as purposive behavior', in HARNAD, S.R., STEKLIS, H.D. and LANCASTER, J. (Eds) *Annals of the New York Academy of Sciences*, 280, pp.212–26.

GLASERSFELD, E.VON (1976b) 'Cybernetics and cognitive development', *Cybernetics Forum*, 8, pp.115–20.

GLASERSFELD, E.VON (1977) 'The Yerkish language and its automatic parser', in RUMBAUGH, D.M. (Ed) *Language learning by a Chimpanzee — The Lana Project*, New York, Academic Press, pp.91–130.

GLASERSFELD, E.VON (1978) 'Radical constructivism and Piaget's concept of knowledge', in MURRAY, F.B. (Ed) *The Impact of Piagetian Theory*, Baltimore, University Park Press, pp.109–22.

GLASERSFELD, E.VON (1979) 'Cybernetics, experience, and the concept of self', in OZER, M.N. (Ed) *A Cybernetic Approach to the Assessment of Children: Toward a More Humane use of Human Beings*, Boulder, Colorado, Westview Press, pp.67–113.

GLASERSFELD, E.VON (1980) 'Viability and the concept of selection', *American Psychologist*, 35, 11, pp.970–74.

GLASERSFELD, E.VON (1981a) 'An attentional model for the conceptual construction of units and number', *Journal for Research in Mathematics Education*, 12, 2, (Reprinted in Steffe *et al.*, 1983), pp.83–94.

GLASERSFELD, E.VON (1981b) 'Feedback, induction, and epistemology', in LASKER, G.E. (Ed) *Applied Systems and Cybernetics*, Vol.2, New York, Pergamon Press, pp.712–19.

GLASERSFELD, E.VON (1981c) 'The conception and perception of number', in GEESLIN, W.E. and WAGNER, S. (Ed) *Models of Mathematical and Cognitive Development*, Columbus, Ohio, ERIC, pp.15–46.

GLASERSFELD, E.VON (1982) 'An interpretation of Piaget's constructivism', *Revue Internationale de Philosophie*, 36, 4, pp.612–35.

GLASERSFELD, E.VON (1983) 'On the concept of interpretation', *Poetics*, 12, pp.207–18.

GLASERSFELD, E.VON (1984) 'An introduction to radical constructivism', in WATZLAWICK, P. (Ed) *The Invented Reality*, New York, Norton, (German edition, Piper, 1981), pp.17–40.

GLASERSFELD, E.VON (1985) 'Reconstructing the concept of knowledge', *Archives de Psychologie*, 53, pp.91–101.

GLASERSFELD, E.VON (1986) 'Steps in the construction of "Others" and "Reality"', in TRAPPL, R. (Ed) *Power, Autonomy, Utopia*, London/New York, Plenum Press, pp.107–16.

GLASERSFELD, E.VON (1987) 'Preliminaries to any theory of representation', in JANVIER, C. (Ed) *Problems of Representation in the Teaching and Learning of Mathematics*, Hillsdale, New Jersey, Lawrence Erlbaum, pp.215–25.

GLASERSFELD, E.VON (1989a) 'Constructivism in education', in HUSEN, T. and

POSTLEWHAITE, T.N. (Eds) *The International Encyclopedia of Education, Supplemental Vol.1*, London/New York, Plenum Press, pp.162–3.

GLASERSFELD, E.VON (1989b) 'Facts and the self from a constructivist point of view', *Poetics*, 18, 4–5, pp.435–48.

GLASERSFELD, E.VON (1992a) 'Warum sprechen wir, und die Schimpansen nicht?' (Why do we speak, and chimpanzees do not?) in HOSP, I. (Ed) *Sprachen der Menschen, Sprache der Dinge*, Bolzano, Italy, Südtiroler Kulturinstitut, pp.53–62.

GLASERSFELD, E.VON (1992b) 'A constructivist view of learning and teaching', in DUIT, R., GOLDBERG, F. and NIEDDERER, H. (Eds) *Research in Physics Learning. Theoretical Issues and Empirical Studies*, Bremen, IPN, pp.29–39.

GLASERSFELD, E.VON (1992c) 'Guest editorial', *Educational Studies in Mathematics*, 23, 3, pp.443–4.

GLASERSFELD, E.VON (1993) 'Notes on the concept of change', in MONTANGERO, J., CORNU, A., TRYPHON, A. and VONÈCHE, J. (Eds) *Conceptions of Change Over Time*, Cahiers de la Fondation Archives Jean Piaget, No.13, Geneva, Archives Jean Piaget, pp.91–6.

GLASERSFELD, E.VON and BARTON BURNS, J. (1962) 'First draft of an English input procedure', *Methodos*, 14, 54, pp.47–79.

GLASERSFELD, E.VON and STEFFE, L.P. (1991) 'Conceptual models in educational research and practice', *Journal of Educational Thought*, 25, 2, pp.91–103.

GOFFMAN, E. (1956) 'The nature of deference and demeanor', *American Anthropologist*, 58, pp.402–73.

GOGUEN, J. (1975) *A Junction Between Computer Science and Category Theory* (Research Report RC 5243), Yorktown Heights, IBM.

GRUBER, H.E. and VONÈCHE, J.J. (Eds) (1977) *The Essential Piaget*, London, Routledge and Kegan Paul.

GUTHRIE, W.K.C. (1962) *The Earlier Presocratics and the Pythagoreans*, Cambridge, Cambridge University Press.

GUTHRIE, W.K.C. (1971) *The Sophists*, Cambridge, Cambridge University Press.

HANSON, N.R. (1958) *Patterns of Discovery*, Cambridge, The University Press.

HEISENBERG, W. (1955) *Das Naturbild der heutigen Physik* (Nature, as seen by today's physics), Hamburg, Rowohlt.

HELMHOLTZ, H.von (1881/1977) *Epistemological Writings* in HERTZ, P. and SCHLICK, M., Dordrech, Holland, Reidel (Originally published 1921).

HERSH, R. (1979) 'Some proposals for reviving the philosophy of mathematics', *Advances in Mathematics*, 31, pp.31–50.

HILGARD, E.R. (1974) 'Toward a neo-dissociation theory: Multiple cognition controls in human functioning', *Perspectives in Biology and Medicine*, 17, 3, pp.301–16.

HUME, D. (1742) *Philosophical Essays Concerning Human Understanding*, London, Millar (in 1758, after the 4th edition, the work was called *An Enquiry Concerning Human Understanding*).

References

HUMBOLDT, W. VON (1907) *Werke*, Vol.7, part 2, Berlin, Leitmann.

HUSSERL, E. (1887/1970) *Philosophie der Arithmetik* (Philosophy of arithmetic), Den Haag, Martinus Nijhoff (first published, 1887).

INHELDER, B. and de CAPRONA, D. (1992) 'Vers le constructivisme psychologique: Structures? Procédures? Les deux indissociables', in INHELDER, B. and CELLÉRIER, G. *Le cheminement des découvertes de l'enfant*, Neuchâtel, Delachaux et Niestlé, pp.19–50.

JAMES, W. (1880) 'Great men, great thoughts, and the environment', *The Atlantic Monthly*, 46, 276, pp.441–59.

JAMES, W. (1912) *Essays in Radical Empiricism*, New York, Longmans, Green.

JAMES, W. (1955) *Pragmatism*, Cleveland/New York, Meridian Books (first published, 1907).

JAMES, W. (1962) *Psychology: Briefer Course*, New York, Collier Books (first published, 1892).

JOAD, C.E.M. (1936) *Guide to Philosophy*, London, Victor Gollancz.

JOYCE, J. (1939) *Finnegans Wake*, London, Faber and Faber.

KAMII, C. and JOSEPH, L.L. (1989) *Young Children Continue to Reinvent Arithmetic, 2nd Grade Implications of Piaget's Theory*, New York, Teachers College Press.

KANT, I. (1781) *Kritik der reinen Vernunft* (Critique of pure reason, 1st edition) in *Kants Werke*, Berlin, Akademieausgabe, Vol.IV.

KANT, I. (1783) *Prolegomena zu einer jeden zukünftigen Metaphysik* (Prolegomena to any future metaphysics) in *Kants Werke*, Berlin, Akademieausgabe, Vol.IV.

KANT, I. (1785) *Grundlegung zur Metaphysik der Sitten* (The metaphysics of morals) in *Kants Werke*, Berlin, Akademieausgabe, Vol.IV, pp.387–463.

KANT, I. (1787) *Kritik der reinen Vernunft* (Critique of pure reason; 2nd edition) in *Kants Werke*, Berlin, Akademieausgabe, Vol.III.

KANT, I. (1798) *Der Streit der Facultäten* (The quarrel of the faculties) in *Kants Werke*, Berlin, Akademieausgabe, Vol.VII, pp.1–116.

KANT, I. (1800) *Anthropologie in pragmatischer Hinsicht* (Anthropology viewed pragmatically) in *Kants Werke*, Berlin, Akademieausgabe, Vol.VII, pp.117–333.

KANT, I. (1803) *Pädagogik* (Pedagogy) in *Kants Werke*, Berlin, Akademieausgabe, Vol.IX, pp.437–499.

KAPUT, J.J. (1991) 'Notations and representations as mediators of constructive processes', in GLASERSFELD E.VON (Ed) *Radical Constructivism in Mathematics Education*, Dordrecht, Kluwer, pp.53–74.

KAUFFMAN, L.H. (1987) 'Self-reference and recursive forms', *Journal of Social and Biological Structure*, 10, pp.53–72.

KEARNEY, R. (Ed) (1985) *The Irish Mind*, Dublin, Wolfhound Press.

KELLY, G. (1963) *A Theory of Personality: The Psychology of Personal Constructs*, New York, Norton.

KÖHLER, W. (1951) 'Untitled contribution to the discussion of a paper by McCulloch at the Hixon Symposium', in JEFFRESS, L.A. (Ed) *Cerebral*

Mechanisms in Behavior, New York, Wiley (reprinted in McCulloch, 1970), pp.42–111.

KUHN, T.S. (1962) *The Structure of Scientific Revolutions*, Chicago, University of Chicago Press, (2nd edition, 1970).

LANGER, S.K. (1948) *Philosophy in a New Key*, New York, Mentor Books.

LASHLEY, K. (1951) 'The problem of serial order in perception', in JEFFRESS, L.A. (Ed) *Cerebral Mechanisms in Behavior*, New York, Wiley.

LAX, W. (1993) 'Symposium on family therapy', *Annual Meeting of the American Society for Cybernetics* (Philadelphia).

LIPSITT, L.P. (1966) 'Learning processes of newborns', *Merrill Palmer Quarterly*, 6, pp.67–76.

LOCHHEAD, J. (1988) 'Some pieces of the puzzle', in FORMAN, G. and PUFFALL, P. (Eds) *Constructivism in the Computer Age*, Hillsdale, New Jersey, Erlbaum, pp.71–81.

LOCKE, J. (1690) *An Essay Concerning Human Understanding*, in FRASER, A.C. (Ed) (1959) New York, Dover.

LORENZ, K. (1979) 'Kommunikation bei Tieren' (Communication in animals), in PEISL, A. and MOHLER, A. (Eds) *Der Mensch uns seine Sprache* (Man and his language), Vienna, Propyläen Verlag, pp.167–80.

LUCE, A.A. and JESSOP, T.E. (Eds) (1950) *The Works of George Berkeley, Bishop of Cloyne*, Vol.I-IX, London, Nelson.

MATTHEWS, M.R. (1992) 'Constructivism and the empiricist legacy', in PEARSALL, M.K. (Ed) *Scope, Sequence, and Coordination of Secondary School Science*, Washington, The National Science Teachers Association, pp.183–96.

MATURANA, H. (1970a) *Biology of Cognition* (Report #9.0), Urbana, Illinois, BCL, University of Illinois.

MATURANA, H.R. (1970b) 'Neurophysiology of cognition', in GARVIN, P.L. (Ed) *Cognition: A Multiple View*, New York, Spartan Books, pp.3–23.

MATURANA, H. (1988) 'Reality: The search for objectivity or the quest for a compelling argument', *The Irish Journal of Psychology*, 9, 1, pp.25–82.

MATURANA, H.R. and VARELA, F.J. (1980) *Autopoiesis and Cognition*, Dordrecht/Boston, Reidel.

MAYR, O. (1970) *The Origin of Feedback Control*, Cambridge, Massachusetts, MIT Press.

McCULLOCH, W.S. (1970) *Embodiments of Mind*, Cambridge, Massachusetts, MIT Press (first published, 1965).

McLELLAN, J.A. and DEWEY, J. (1908) *The Psychology of Number*, New York, Appleton.

McMULLIN, E. (Ed) (1988) *Construction and Constraint*, Notre Dame, Indiana, University of Notre Dame Press.

MEDAWAR, P. (1984) *The Limits of Science*, Oxford, Oxford University Press.

MEYENDORFF, J. (1974) *Byzantine Theology*, New York, Fordham University Press.

MINSTREL, J. (1992) 'Facets of students' knowledge and relevant instruction', in Duit, R., Goldberg, F. and Niedderer, H. (Eds) *Research in Physics Learning — Theoretical Issues and Empirical Studies*, Kiel, Germany, IPM, pp.110–28.

Moessinger, P. and Poulin-Dubois, D. (1981) 'Piaget on abstraction', *Human Development*, 24, pp.347–53.

Nash, J. (1970) *Developmental Psychology*, Englewood Cliffs, New Jersey, Prentice-Hall.

Ogden, C.K. (Ed) (1959) *Bentham's Theory of Fictions*, Paterson, New Jersey, Littlefield, Adams and Co. (First published, London, Routledge and Kegan Paul, 1932.)

Pask, G. (1961) *An Approach to Cybernetics*, New York, Harper and Brothers.

Peano, G. (1891a) 'Principii di logica matematica', in Peano, G. (Ed) (1891) *Rivista di matematica*, Turin, Bocca.

Peano, G. (1891b) 'Sul concetto di numero', in Peano, G. (Ed) (1891) *Rivista di matematica*, Turin, Bocca.

Piaget, J. (1929) 'Les deux directions de la pensée scientifique' (The two directions of scientific thought), *Archives des Sciences Physiques et Naturelles*, 11, pp.145–65.

Piaget, J. (1937) *La construction du réel chez l'enfant* (The construction of reality in the child, Translation, M.Cook, New York, Basic Books, 1971), Neuchâtel, Delachaux et Niestlé.

Piaget, J. (1945) *La formation du symbole chez l'enfant* (Play, dream, and imitation in childhood), Neuchâtel, Delachaux et Niestlé.

Piaget, J. (1952a) *The Child's Conception of Number* (Translation, Gattegno and Hodgson), London, Rouledge and Kegan Paul (French original, 1941).

Piaget, J. (1952b) 'Jean Piaget', in Boring, E., Langfeld, H., Werner, H. and Yerkes, R. (Eds) *A History of Psychology in Biography*, Vol.4, Worcester, Massachusetts, Clark University Press, pp.237–56.

Piaget, J. (1957) *Logic and Psychology* (translation, W.Mays), New York, Basic Books.

Piaget, J. (1965) *Études sociologiques* (Sociological studies), Geneva, Librairie Droz (Italian translation by Barbetta, P. and Fornasa, W. 1989, Milan, Franco Angeli).

Piaget, J. (1967a) *Biologie et connaissance* (Biology and knowledge), Paris, Gallimard.

Piaget, J. (1967b) *Six Psychological Studies*, New York, Vintage. (French original: Geneva, 1964)

Piaget, J. (1967c) 'Le système et la classification des sciences' (System and classification of the sciences), in Piaget, J. (Ed) *Logique et connaissance scientifique*, Paris, Encyclopédie de la Pléiade, Gallimard, pp.1151–1224.

Piaget, J. (1968) *On the Development of Memory and Identity* (Translation by E. Duckworth), Worcester, Massachusetts, Clarke University Press.

Piaget, J. (1969) *The Mechanisms of Perception* (Translation Seagrim), New York, Basic Books (French original, 1961).

Piaget, J. (1970a) *Genetic Epistemology*, New York, Columbia University Press.

PIAGET, J. (1970b) *Le structuralisme* (Structuralism), Paris, Presses Universitaires de France, 4th edition.

PIAGET, J. (1970c) *Main Trends in Interdisciplinary Research*, New York, Harper Torchbooks.

PIAGET, J. (1974a) *La prise de conscience* (The grasp of consciousness), Paris, Presses Universitaires de France.

PIAGET, J. (1974b) *Réussir et comprendre* (Success and understanding), Paris, Presses Universitaires de France.

PIAGET, J. (1974c) *Adaptation vitale et psychologie de l'intelligence* (Adaptation and intelligence), Paris, Hermann.

PIAGET, J. (1975) *L'équilibration des structures cognitives* (The equilibration of cognitive structures), Paris, Presses Universitaires de France.

PIAGET, J. (1976a) 'Piaget's Theory', in INHELDER, B. and CHIPMAN, H.H. (Eds) *Piaget and His School*, New York, Springer (Originally published in Mussen, P.H. (Ed) *Carmichael's Manual of Child Psychology*, Vol.1, New York, Wiley (1970), pp.703–710).

PIAGET, J. (1976b) 'Autobiographie, partie ix (1966–1976)', (Autobiography, part ix), *Revue européenne des sciences sociales*, 14, pp.35–43.

PIAGET, J. (1977a, *et al.*) *Recherches sur l'abstraction réfléchissante*, Vol.1 and 2 (Research on reflective abstraction), Paris, Presses Universitaires de France.

PIAGET, J. (1977b) 'Appendix B', in INHELDER, B., GARCIA, R. and VONÈCHE, J. *Épistémologie génétique et équilibration*, Neuchâtel, Delachaux et Niestlé, pp.90–2.

PIAGET, J. and GARCIA, R. (1983) *Psychogénèse et histoire des sciences*, Paris, Flammarion.

PIRIE, S. and KIEREN, T. (1989) 'A recursive theory of mathematical understanding', *For the Learning of Mathematics*, 9, 3, pp.7–11.

PISANI, P. (1977) 'Computer programs', in RUMBAUGH, D.M. (Ed) *Language Learning by a Chimpanzee — The Lana Project*, New York, Academic Press, pp.131–42.

PLATO (ca. 390 BC) 'The republic', in WARMINGTON, E.H. and ROUSE, P.G. (Eds) (1956) *Great Dialogues of Plato*, New York, Mentor.

POINCARÉ, H. (1952) *Science and Hypothesis* (translation, W.J.G.), New York, Dover (French original, 1902.)

POLLACK, R.H. and BRENNER, M.W. (1969) *The Experimental Psychology of Alfred Binet*, New York, Springer.

POPKIN, R.H. (1979) *The History of Scepticism from Erasmus to Spinoza*, Berkeley, California, University of California Press.

POPPER, K. (1968) *Conjectures and Refutations: The Growth of Scientific Knowledge*, New York, Harper Torchbooks.

POWERS, W.T. (1973) *Behavior: The Control of Perception*, Chicago, Aldine.

POWERS, W.T. (1978) 'Quantitative analysis of purposive systems: Some spadework at the foundations of scientific psychology', *Psycholocical Review*, 85, 5, pp.417–35.

PRITCHARD, R.M., HERON, W. and HEBB, D.O. (1960) 'Visual perception approached by the method of stabilized images', *Canadian Journal of Psychology*, 14, 2, pp.67–77.

PUTNAM, H. (1981) *Reason, Truth and History*, Cambridge, Cambridge University Press.

RENSHAW, P.D. (1992) 'The psychology of learning and small group work', in MCLEAN, R. (Ed) *Classroom Oral Language*, (pp.90–94) Deakin, Australia, Deakin University Press, pp.90–4.

RIEBER, L.P. (1993) 'A pragmatic view of instructional technology', in TOBIN, K. (Ed) *The Practice of Constructivism in Science Education*, Washington, AAAS Press, pp.193–212.

RORTY, R. (1989) *Contingency, Irony, and Solidarity*, Cambridge, Cambridge University Press.

ROSENBLATT, L.M. (1985) 'Viewpoints: Transaction versus interaction — A terminological rescue operation', *Research in Teaching English*, 19, 1, pp.98–107.

ROSSI-LANDI, F. (1961) *Significato, comunicazione e parlare comune* (Meaning, communication, and ordinary language), Padua, Marsilio.

ROTENSTREICH, N. (1974) 'Humboldt's prolegomena to philosophy of language', *Cultural Hermeneutics*, 2, pp.211–27.

RUSSELL, B. (1956) 'Definition of number', in NEWMAN, J.R. (Ed) *The World of Mathematics*, New York, Simon and Schuster, pp.529–34.

RUSSELL, B. (1986) *Mysticism and Logic*, London, Unwin Paperbacks (first published, 1917).

RUSSELL, S.J. and CORWIN, R. (1991) 'Talking mathematics: "Going slow" and "Letting go"', *Proceedings of the 13th Annual Meeting of the North American Chapter of PME*, Vol.2, pp.175–81.

SAPIR, E. (1921) *Language*, New York, Harcourt, Brace and World.

SAUSSURE, F. DE (1959) *Course on General Linguistics* (Translation by W. Baskin), New York, Philosophical Library (French original, 1916).

SCHOPENHAUER, A. (1819) *Die Welt als Wille und Vorstellung* (The world as will and idea), Stuttgart, Cotta'sche Buchhandlung.

SECORD, P.F. and PEEVERS, B.H. (1974) 'The development and attribution of person concepts', in MISCHEL, T. (Ed) *Understanding Other Persons*, Oxford, Basil Blackwell.

SHANNON, C.E. (1948) 'The mathematical theory of communication', *Bell Systems Technical Journal*, 27, pp.379–423 and 623–56.

SHAW, G.B. (1923) *Saint Joan*.

SIMMEL, G. (1895) 'Über eine Beziehung der Selectionslehre zur Erkenntnistheorie' (About a relation between the doctrine of natural selection and the theory of knowledge), *Archiv für systematische Philosophie*, 1, pp.34–45.

SINCLAIR, H. (1990) 'Learning: The interactive recreation of knowledge', in STEFFE, L.P. and WOOD, T. (Eds) *Transforming Children's Mathematics Education*, Hillsdale, New Jersey, Erlbaum, pp.19–29.

SKINNER, B.F. (1971) *Beyond Freedom and Dignity*, New York, Bantam Books.

SKINNER, B.F. (1977) 'Why I am not a cognitive psychologist', *Behaviorism*, 5, 2, pp.1–10.

SMITH, L. (1981) 'Piaget mistranslated', *Bulletin of the British Psychological Society*, 4, pp.1–3.

SMOCK, C.D. and GLASERSFELD, E.VON (1974) *Epistemology and Education: The Implications of Radical Constructivism for Knowledge Acquisition* (Report #14), Athens, Georgia, Follow Through Publications.

SOMENZI, V. (1969) *La fisica della mente* (The physics of the mind), Turin, Boringhieri.

SOMENZI, V. (1991) *La materia pensante* (Matter that thinks), Milan, CLUP.

SPENCER BROWN, G. (1973) *Laws of Form*, New York, Bantam Books (First published, 1969)

STEFFE, L.P. (1984) 'Children's construction of number sequences and multiplying schemes', in HIEBERT, J. and BEHR, M. (Eds) *Number Concepts in the Middle Grades*, Hillsdale, New Jersey, Erlbaum, pp.119–46.

STEFFE, L.P. (1991) 'The constructivist teaching experiment: Illustrations and implications', in GLASERSFELD, E.VON (Ed) *Radical Constructivism in Mathematics Education* Dordrecht, Kluwer, pp.177–94.

STEFFE, L.P. (Ed) (1991) *Epistemological Foundations of Mathematical Experience*, New York, Springer Verlag.

STEFFE, L.P. and COBB, P. (1988) *Construction of Arithmetical Meanings and Strategies*, New York, Springer.

STEFFE, L.P. and GLASERSFELD, E.VON (1988) 'On the construction of the counting scheme', in STEFFE, L.P. and COBB, P., pp.1–19.

STEFFE, L.P., GLASERSFELD, E.VON, RICHARDS, J. and COBB. P. (1983) *Children's Counting Schemes: Philosophy, Theory, and Application*, New York, Praeger Scientific.

STEFFE, L.P., RICHARDS, J. and GLASERSFELD, E.VON (1978) 'Experimental models for the child's acquisition of counting and of addition and subtraction', in FUSON, K. and GEESLIN, W.E. (Eds) *Explorations in the Modeling of the Learning of Mathematics* Columbus, Ohio, ERIC/SMEAC Center for Science, Mathematics, and Environmental Education, pp.27–44.

STEFFE, L.P., THOMPSON, P. and RICHARDS, J. (1982) 'Children's counting in arithmetical problem solving', in CARPENTER, T., MOSER, J. and ROMBERG, T. (Eds) *Addition and Subtraction: A Cognitive Perspective*, Hillsdale, New Jersey, Erlbaum, pp.211–23.

THORNDIKE, E.L. (1931) *Human Learning*, New York, Century.

TOMASELLO, M. (1992) *First Verbs: A Case Study of Early Grammatical Development*, Cambridge, Cambridge University Press.

TYMOCZKO, T. (1994) 'Humanistic and utilitarian aspects of mathematics', in ROBITAILLE, D.E., WHEELER, D.H. and KIERAN, C. (Eds) *Selected Lectures From the 7th International Congress on Mathematics Education*, Sainte-Foi, Canada, Presses de l'Université Laval, pp.327–39.

UEXKÜLL, J.VON and KRISZAT, G. (1933) *Streifzüge durch die Umwelten von Tieren und Menschen*, Frankfurt am Main, Fischer.

VACCARINO, G. (1977) *La chimica della mente* (The mind's chemistry), Messina, Italy, Carbone.

VACCARINO, G. (1981) *Analisi dei significati* (Analysis of meanings), Rome, Italy, Armando.

VACCARINO, G. (1988) *Scienza e semantica costruttivista* (Science and constructivist semantics), Milan, Italy, CLUP.

VAIHINGER, H. (1913) *Die Philosophie des Als Ob* (The philosophy of as if), Berlin, Reuther and Reichard, 2nd edition.

VALÉRY, P. (1957) *Œuvres* (Works), Paris, Bibliothèque de la Pléiade.

VARELA, F.J. (1975) 'A calculus for self-reference', *International Journal of General Systems*, 2, pp.5–24.

VICO, G-B. (1710) *De antiquissima Italorum sapientia* (Latin original and Italian translation by Pomodoro, 1858), Naples, Stamperia de'Classici Latini.

VICO, G-B. (1744) *Principi di scienza nuova* (The New Science; Translation by BERGIN, T.G. and FISCH, M.H., 1961), Garden City, New York, Anchor Books.

VONÉCHE, J. (1982) 'Evolution, development, and the growth of knowledge'. In BROUGHTON, J.M. and FREEMAN-MOIR, D.J. (Eds) *The Cognitive Developmental Psychology of James Mark Baldwin*, Norwood, New Jersey, Ablex, pp.51–79.

VUYK, R. (1981) *Overview and Critique of Piaget's Genetic Epistemology*, Vol.1 and 2, London/New York, Academic Press.

VYGOTSKY, L.S. (1962) *Thought and Language*, Cambridge, Massachusetts, MIT Press.

WATZLAWICK, P. (1977) *How Real is Real?*, New York, Vintage.

WATZLAWICK, P. (1984) *The Invented Reality*, New York, Norton (German original, München: Piper, 1981).

WHORF, B.L. (1956) *Language, Thought, and Reality*, Cambridge, Massachusetts, MIT Press.

WIENER, N. (1965) *Cybernetics*, Cambridge, Massachusetts, MIT Press (first published 1948).

WITTGENSTEIN, L. (1933) *Tractatus Logico-philosophicus*, London, Kegan-Paul, Trench, Trubner and Co.

WITTGENSTEIN, L. (1953) *Philosophical Investigations*, Oxford, Basil Blackwell.

WITTGENSTEIN, L. (1965) *The Blue and Brown Books*, New York, Harper Torchbooks (first published 1958).

WITTGENSTEIN, L. (1969) *On Certainty*, Oxford, Basil Blackwell.

WOOD, T., COBB, P. and YACKEL, E. (1993) *Purdue Problem Centered Mathematics Project: Summary*, West Lafayette, Indiana, School of Education, Purdue University.

ZAZZO, R. (1979) 'Des enfants, des singes et des chiens devant le miroir' (Of children, monkeys, and dogs in front of the mirror), *Revue de Psychologie Appliquée*, 20, 2, pp.235–46.

ZINCHENKO, V.P. and VERGILES, N.Y. (1972) *Formation of Visual Images, Special Research Report*, New York, Consultants Bureau.

Index of Names

Accame, Felice 23, 175
Achilles 185, 186
Albani, Enrico 23
Amherst (Massachusetts, US) 20
Ampère, André-Marie 147
Aristotle 108
Ashby, Ross W. 148
Athens (Georgia, US) 9

Bacon, Francis 161
Baldwin, James Mark 74
Balzac, Honoré de 3
Barbetta, Pietro 128
Barthes, Roland v, 83
Barton, Jehane 9
Bateson, Gregory 18, 46, 148, 155–6, 197
Bauersfeld, Heinrich 183, 190, 191
Beer, Stafford 148
Bellarmino (Cardinal) 29, 51
Belloni, Lanfranco 29, 51
Bentham, Jeremy 36, 45, 52
Berger, Peter 122–3
Berkeley, George (Bishop of Cloyne) 4–5, 31, 33–4, 35, 36, 41, 49, 51, 52, 91, 92, 93, 96, 97, 122, 164–5
Bernal, J.D. 42
Bernard, Claude 146
Beth, Evert W. 69
Bickhard, Mark 115
Binet, Alfred 10
Bogdanov, Alexander 120–21
Bohr, Niels 23, 149
Born, Max 23
Bower, T.G.R. 84
Boyle, C.F. 189
Bridgman, Percy W. 6–7, 23, 76, 78, 79, 160, 164
Bringuier, Jean-Claude 71
Bronowski, Jacob 25
Broughton, J.M. 74
Brouwer, Luitzen E.J. 23, 46
Byzantium 27

Campbell, Donald 44
Campbell, Robert L.
Caprona, Denys de 64
Caramuel, Juan (Bishop of Vigevano) 108, 112, 163–4, 166, 170
Carpenter, Ray 11
Ceccato, Silvio 6, 7, 12, 76, 78, 79, 80, 88, 97, 111, 167, 175
Cellérier, Guy 67
Cherry, Colin 7
Chomsky, Noam 15, 139
Clement, John 184, 187
Cobb, Paul 16, 100, 137, 183, 190
Collingwood, Robin G. 25
Conant, Roger 148
Confrey, Jere 188
Copernicus, Nicolas 29
Corwin, R. 188
Craik, Kenneth 152

Darwin, Charles 42, 43, 50
Désautels, Jacques 177
Descartes, René 28, 30–31, 32, 36, 49, 122
Dewey, John 25, 89, 90, 164
Diels, Hermann 51
Dirac, Paul 23
diSessa, Andy 187
Driver, Rosalind 187
Dublin 4–6
Duckworth, Eleanor 12
Duhem, Pierre 46
Dutton, Brian 9
Dykstra, Dewey 189

Einstein, Albert 6, 21, 22, 23, 149, 165
Eriugena, John Scottus 28, 39, 49
Ernest, Paul xv
Euclid 16, 162, 163, 169, 171
Exeter xii
Exner, Franz 52

Index of Names

Whorf, Benjamin L. 3
Wiener, Norbert 147, 148, 157
Wittgenstein, Ludwig 3, 4, 10, 87, 122, 123, 133, 134, 140
Wood, Terry 183, 190

Xenophanes 26

Yackel, Erna 194, 204

Zazzo, René 126
Zeno of Elea 80, 185, 186
Zinchenko, V.P. 10, 116, 167
Zuoz (Switzerland) 2
Zürich 3

Subject Index

Made in the USA
Lexington, KY
19 January 2016